A TALE OF SEVEN SCIENTISTS AND
A NEW PHILOSOPHY OF SCIENCE

A TALE OF SEVEN SCIENTISTS AND A NEW PHILOSOPHY OF SCIENCE

Eric Scerri

OXFORD
UNIVERSITY PRESS

OXFORD
UNIVERSITY PRESS

Oxford University Press is a department of the University of Oxford.
It furthers the University's objective of excellence in research, scholarship,
and education by publishing worldwide. Oxford is a registered trade mark of
Oxford University Press in the UK and certain other countries.

Published in the United States of America by Oxford University Press
198 Madison Avenue, New York, NY 10016, United States of America.

© Eric Scerri 2016

Library of Congress Cataloging-in-Publication Data
Names: Scerri, Eric R.
Title: A tale of seven scientists and a new philosophy of science/Eric Scerri.
Description: New York, NY: Oxford University Press, 2016. |
Includes bibliographical references and index.
Identifiers: LCCN 2016006032 | ISBN 9780190232993
Subjects: LCSH: Science—Philosophy. | Philosophy and science.
Classification: LCC Q175 .S3044 2016 | DDC 501—dc23 LC record available at
http://lccn.loc.gov/2016006032

1 3 5 7 9 8 6 4 2
Printed by Sheridan Books, Inc., United States of America

ACKNOWLEDGMENTS

This book is dedicated to my mother, Ines Scerri, who passed away in April of 2015, and to my wife Elisa for her love and encouragement.

I would also like to thank the following colleagues and friends who have had an influence on my views or have made direct comments on the material that I discuss. My apologies to anyone whose name I might have inadvertently omitted.

Micheal Akeroyd, Roger Alberto, Roy Alexander, Rolando Alfaro, Santiago Alvarez, Theodore Arabatzis, Waldmir Arujo-Neto, Peter Atkins, Philip Ball, Marina Banchetti, Anne-Sophie Barwich, Karim Bschir, Louis Bouchard, Stuart Cantrill, Loic Casson, Alan Chalmers, Jose Antonio Chamizo, Hasok Chang, Anjan Chakravartty, Oliver Choroba, Simon Cotton, Adrian Dingle, Alexander Ehmann, Grant Fisher, Dan Garber, Elena Ghibaudi, Amihud Gilead, Carmen Giunta, Michael Gordin, Brian Gregory, Steve Hardinger, Rom Harre, Chris Haufe, Clevis Headley, John Heilbron, Hinne Hettema, Richard Hirsh, Roald Hoffman, Ann Hong-Hermesdorf, Bill Jensen, Ric Kaner, Jouni-Matti Kuukkanen, Gregg Kallo, Masanori Kaji, Helge Kragh, Martin Labarca, David Lamb, Mark Leach, Jeffrey Leigh, Lucia Lewowicz, Olimpia Lombardi, Jean-Pierre Llored, Steve

Lyons, Tim Lyons, Farsad Mahootian, Alex Manafu, James Marcum, Tom Mason, Michael Matthews, Russell McCorrmach, Chas McCaw, Lee McIntyre, Fabienne Meyers, Peter Morris, Lars Öhrström Yuri Orlik, Tina Overton, Sergio Palazzi, Will Parsons, Raffaele Pisano, Vincenzo Politi, Martyn Poliakoff, Ted Porter, Pekka Pyykko, Geoffrey Rayner Canham, Guillermo Restrepo, Marcos Ribeiro, Alan Rocke, James Rota, Klaus Ruthenberg, Sam Schindler, Raphael Scholl, Eugen Schwarz, Jeffrey Seeman, Sylvain Smadja, Betty Smokovitis, Kyle Stanford, Philip Stewart, Keith Taber, Marco Taddia, Jess Tauber, Pieter Thyssen, John Torday, Valery Tsimmerman, Rene Vernon, Jose Luis Villaveces, Peter Vickers, Donny Wise, Norton Wise, Brad Wray, Alfio Zambon.

Special thanks to my editor at OUP, Jeremy Lewis, as well as all the editorial staff including Gwen Colvin, Julie Fergus, and Anna Langley.

Finally, thanks to Bruce Main Smith who provided a photo of his father John David Main Smith and to Abigail Bury for providing a photo of her grandfather, Charles Bury.

CONTENTS

Foreword by Peter Atkins   ix
Foreword by James A. Marcum   xi
Biographical Background   xv

Chapter 1 Introduction   3

Chapter 2 John Nicholson   13

Chapter 3 Anton Van den Broek   41

Chapter 4 Richard Abegg   63

Chapter 5 Charles Bury   79

Chapter 6 John D. Main Smith   103

Chapter 7 Edmund Stoner   117

Chapter 8 Charles Janet   149

Chapter 9 Bringing Things Together   171

Index   219

FOREWORD

The periodic table is a deep pool of fascination. There is of course its everyday utility, which is perhaps more relevant to the education of chemists than the pursuit of research. Instructors everywhere (like Mendeleev himself) find it indispensable for ordering instruction and enlightenment. There is also the titillation of pressing at its frontiers to reveal new elements and the satisfaction of completing a row, however useless the newly created elements almost certainly will prove to be.

Eric Scerri has been exploring the foundations of the periodic table for many years and has presented his findings in a number of books. In this one he shifts his attention from mainly chemists to mainly physicists and migrates to the early 20th century when the structure of the atom became clear and periodicity was finally understood. He focuses on what he terms the "little people," the underpinnings of progress, such as Nicholson, Bury, Abegg, Main Smith, and Stoner. They are largely forgotten today yet in their time made pivotal contributions to our knowledge of how electrons are distributed around the nucleus of an atom. Matter, and specifically its chemical building blocks the elements, is so varied, and until its atomic structure had been elucidated, so perplexing that it is not surprising that so many—both chemists and physicists—contributed to attempts to understand it.

One strength of science is that it is often wrong. It proceeds by overthrowing preconception, perhaps replacing misconception by a more sophisticated misconception until that misconception is replaced by yet another, until ultimately (we optimists all hope) arriving at some version of the truth. The underlying explanation for the periodic table shows how intellectual order can emerge from chaos, error, persistence, and the successive overthrow of misconception and is, more or less, now "true." As such its development epitomizes the progress of science. As readers will see, Scerri explores this aspect of science in the context of the atomic structure, showing how science typically muddles through rather than progressing like a screenplay by Euclid. For Scerri, science is evolution, not revolution. Indeed, he explores the view that science progresses just like organic evolution, complete with the random mutation of ideas, some of which survive while others simply wither away.

No scientific discovery is an island. All discoveries involve enrichment by interaction with neighboring fields, disputes about priority, collaboration of disciplines, and progress through the weeding out of error. Both historians of science and its philosophers will find much to stimulate them in this book, and science educators will be greatly enriched by learning more about a topic so central to their instruction.

Peter Atkins

Oxford 2016

FOREWORD

Eric Scerri's book on the philosophy of science is a spirited alternative to both traditional analytical and historical philosophy of science. Specifically, Scerri advocates an evolutionary philosophy of science, which challenges Karl Popper's and especially Thomas Kuhn's earlier revolutionary philosophy of science. Scerri constructs this alternative philosophy on the notion of Darwinian gradualism, which includes the systematic advance of scientific knowledge through both the "major" and "minor" figures in science. To that end, he reconstructs the practice of seven scientists, including Richard Abegg, Anton van den Broek, Edmund Stoner, and Charles Janet, as they went about investigating phenomena and topics such as atomic number, chemical bonding, and the periodic table. Moreover, in contrast to the formal logic of analytical philosophy of science Scerri advocates a role for intuitions and instincts as scientists methodically explore the natural world.

Importantly, Scerri's evolutionary philosophy of science is both holistic and organic. It is holistic in the sense that there is a fundamental unity to scientific practice in terms of a community of practitioners, both major and minor figures, who investigate the world and

formulate theories to explain it. His holism is wide-ranging concep-
tually in that it provides a larger picture with respect to the growth of
scientific knowledge in contrast to the anemic traditional view that
privileges only the major figures. From his evolutionary perspective,
the minor figures serve as "missing links" to provide a continuous
account for the growth of scientific knowledge. The end result is a
comprehensive view of science as compared to the narrowly trun-
cated and distorted view that champions only heroic major figures.
Additionally, his evolutionary philosophy of science is organic in that
he compares science to a living organism, so that scientific knowledge unfolds as it is rooted in the activities of scientists who struggle
to understand the world and not in terms of a particular end point
like the truth of theories. Rather, theories are useful adaptations that
allow scientists to piece together slowly the natural world in an intel-
ligible fashion.

Although Scerri's evolutionary philosophy of science seems at
first glance prosaic, as he admits, it is upon further reflection rather
radical, as he also admits. It is prosaic in the sense that the activities of
many scientists seem uninspiring in terms of the monumental growth
of scientific knowledge. Indeed, often the theories proposed by these
scientists are "wrong" compared to the revolutionary theories proposed by the enlightened and brilliant few major figures who are celebrated for profoundly altering the course of science and marshaling
in a new era or paradigm for practicing it. For a traditional philosophy
of science, whether analytical or historical, envisions scientific progress as revolutionary in terms of the competition between right and
wrong theories—with the right theory supplanting the wrong theory. It is here that Scerri's evolutionary philosophy of science is radical in the sense that scientific advancement is possible even when the
theory is wrong. He offers as an example John Nicholson, who was
able to explain particular spectral phenomena and thereby advance
scientific knowledge even though he operated from a "wrong" theory.

In conclusion, Scerri champions an evolutionary philosophy of science that takes seriously the holistic and organic nature of the scientific enterprise from a traditional Darwinian perspective. It is holistic in that he incorporates the contribution of everyone within the scientific community in the march of science to investigate and explain natural phenomena. But more importantly, it is organic, as Scerri writes:

> My account is more organic and less isolationist, more guided by blind chance and evolutionary forces than by human rationality. Seen from a far distance we might even suppose that science is developing as one large interconnected organism. (chapter 9)

In the end, what is great about science is not a few towering figures but the community of practitioners who struggle to make sense of the world. And, that struggle continues to evolve—for the science of today is vastly different from the science of generations past. Finally, in the following pages, Scerri provides the reader with a vision of an evolving science that is both dynamic and inspiring.

James A. Marcum

Baylor University, 2016

BIOGRAPHICAL BACKGROUND

I discovered chemistry and physics at the age of about 13 or 14 while a student at Walpole Grammar School in West London in the mid 1960s. Before that I had been especially interested in history and geography. I can still vividly remember my first physics exam in which we were asked to explain the difference between speed and velocity. Because we had an awful physics teacher we were not taught the difference between scalars and vectors and so I had no clue as to what was being asked. I can recall being especially perplexed by what seemed to me to be two perfectly equivalent ways of saying the same thing.

Before long I became hooked on science and especially on trying to really understand things. My classmates were sometimes puzzled by my insistence on wanting to understand the material, while they seemed to be perfectly happy to solve numerical problems and to move onto the next question. Perhaps the philosophical "disease" was there already, but if so I knew nothing about it and high schools in the British education system did not include any study of philosophy.

For my undergraduate work I gravitated toward the more theoretical parts of chemistry and went directly up to Cambridge, to do a research project in theoretical chemistry. I worked on the magnetic quadrupole moments of small molecules, which was a little formal for my liking. At one point my advisor suggested that I take what he regarded as my philosophical ideas to the department of history and philosophy of science.[1] It was the first time that I had really heard of this academic discipline and the time was not right for me to jump ship completely. Instead I took a year off college. I then took up another project in experimental physical chemistry, which involved work in laser Raman spectroscopy of polymeric molecules, which provided some valuable practical experience. Next, I left formal academic work for a period of about six years during which time I taught high school chemistry in private tutorial colleges in London. The students were generally very good and my knowledge of science grew rapidly. As the saying goes, "One only begins to understand a subject when one has to teach it to others." During this period I had casual encounters with the history and philosophy of science, such as when a student lent me a copy of Alan Chalmers's book, *What Is This Thing Called Science?* I was especially taken by the account of Karl Popper's work and was nothing short of devastated by the problem of induction and what this implied for scientific knowledge—about which I had always had a rather naïve, or perhaps just a typical scientist's unexamined, view about.

In the mid 1980s I happened to be living in South Kensington close to Chelsea College (then part of the University of London). I began hanging out in the library and dipping into books on the fundamentals of chemistry and physics. I soon applied to this school to do a PhD in history and philosophy of science. I originally wanted to do a project in quantum mechanics or relativity theory but was quickly convinced by my future supervisor, Heinz Post, that my background was far better suited to working in the philosophy of chemistry—then a largely uncharted area. This I did from 1986 until 1992, when I emerged with what was probably the first doctorate in this new field of study.

My work consisted of a rather technical investigation into the relationship between chemistry in general (and the periodic table in particular) and quantum mechanics. Of course I studied a good deal of logic, philosophy, and history of science but my emphasis was always on the science rather than the history or the philosophy. In 1995, I embarked on a postdoctoral fellowship in the history of science at the California Institute of Technology but remained very interested in philosophy of chemistry and did all I could to promote the new field. For example I founded, and am still the editor of, the journal *Foundations of Chemistry*.[2]

Let me return briefly to Popper. It has taken me a good deal of time to recover from the initial shock of learning the limitations of inductive science and the general influence of the great man, Popper, whom my advisor had known personally. Although Popper had retired from conferences and other academic occasions, his spirit still lurked in the philosophy of science scene in London.[3] The work of Thomas Kuhn that I was also exposed to at an early stage was immensely appealing because it took a more naturalistic approach and consisted of an examination of the history of science rather than concentrating on the logical issues. However, the social climate in my own department and among "respectable" philosophers of science in London was that Kuhn was responsible for most of the ills in the discipline starting with Shapin and Schaffer's book, *Leviathan and the Air Pump*, the sociological and relativistic turn and the eventual emergence of Science Wars. It is only recently while starting to prepare the present book that I have finally shrugged myself free of the British preference for Popper over Kuhn, and have begun to really appreciate the full value of some of Kuhn's contributions. Nevertheless, as I will be arguing, I still believe that Kuhn is incorrect to emphasize the role of revolutions in science.

One thing that has always remained with me is a general aversion toward analytical philosophy and the role of philosophy of language in modern philosophy of science. And yet all my writings remained in the realm of scientific details and I was never tempted to go via the sociological-anthropological route to the study of science. I also began writing a number of semipopular books on the periodic table and on the elements, in which I examined the role of controversies, priority conflicts, and simultaneous discoveries in 19th- and 20th-century chemistry and physics. In addition I have always tried to rehabilitate the roles of what I call minor players in the history of science, meaning people like van den Broek and Edmund Stoner in physics as well as Charles Bury and John Main Smith in chemistry.[4]

The current book attempts to draw together various disparate strands in my thinking. I have arrived at a point at which I favor an evolutionary approach to the philosophy of science in a rather literal biological sense. In this book I will be defending an evolutionary view of the manner in which science develops. Some of this will bring me into sharp disagreement with the views of Kuhn. In other respects I owe a great debt to Kuhn for having clarified my own ideas on the nature of truth in science. I arrived at what for me was a surprising conclusion, that science evolves from within, as it were, and does not aim at an external objective truth—and then I discovered that Kuhn was already discussing such a view in the final two or three pages of his *Structure of Scientific Revolutions*.[5] But whereas Kuhn seems to believe that science proceeds in a revolutionary fashion while at the same time claiming to embrace an evolutionary epistemology, my view is that one cannot claim both together at the same time.

My own evolutionary view of the growth of science is at the same time both obvious and radical. I take courage from another well-known adage, whereby good philosophy consists of noticing what is obvious to other people. I hope this may be true in my case too and that I may indeed have noticed something obvious that is worth pausing to elaborate. I will be claiming that scientific theories are seldom or ever right or wrong, and that the role of language is not all that it is claimed to be. I believe that intuition, instinct, and urges on the part of scientists are just as important as cold hard logic. I suggest that language and careful analysis comes at a later stage in the evolution of scientific ideas and that half-baked ideas published by virtually unknown scientists such as those I have already named may have been far more influential when viewed from a holistic perspective. I will present a picture of science in which individuals, and especially the traditional giants of the field, are no more important than the minor players. I will paint a picture of science as a far more chaotic activity in which ideas are proposed and continuously modified by other scientists. It is this constant honing that I claim is mostly responsible for what in hindsight constitutes scientific progress.

I will go as far as to suggest that, from this bird's eye or objective perspective that avoids deifying the heroes, the development of science should be regarded as one organic flow in which the individual worker bees are all contributing to the good of the "hive." As I said, the main thesis is both obvious and radical at the same time: Science has a life of its own. It takes place regardless of the wishes and aspirations of individual scientists and is far less governed by logic and rationality than popularly supposed. I appreciate that many readers will find this proposal difficult to accept. But let me recall the fact that scientific progress has often led to a more humble view of the central role that humans are supposed to occupy according to the modernist view. One need only consider the demise of the geocentric model of the solar system or Darwin teaching us that we have evolved from "lower" forms of animals. For a more contemporary example think of how modern astronomers have gone from categorically denying the possibility of there being other planetary systems to now having solid evidence for well over one hundred exoplanets as they are called.[6]

Similarly, some still cannot bring themselves to believe that life might exist on astronomical bodies other than the earth even though the evidence concerning planets beyond the solar system would seem to suggest otherwise. The more scientific knowledge develops, the more our insignificance unfolds itself. Could it be that even that most valued of human achievements, namely the scientific method, might turn out to be something we should not take quite so much credit for if, as I am suggesting, it proceeds almost of its own accord?

OUP BLOG

The following blog was written by me and first appeared online in 2015. It is reproduced here with slight modifications because it provides a quick summary of some of the main ideas that are presented in this book.[7]

One of the central concepts in chemistry is the electronic configuration of atoms. This is equally true of chemical education and in professional chemistry and research. If one knows how the electrons in an atom are arranged, especially in the outermost shells, one immediately understands many properties of an atom, such as how it bonds and its position in the periodic table. I have spent the past couple of years looking closely at the historical development of this concept as it unfolded at the start of the 20th century.

What I have found has led me to propose a new view for how science develops, a philosophy of science if you will, in the grand tradition of attempting to explain what science really is. I am well aware that such projects have fallen out of fashion since the ingenious, but ultimately flawed, attempts by the likes of Popper, Kuhn, and Imre Lakatos in the 1960s and '70s. But I cannot resist this temptation since I think I may have found something that is paradoxically obvious and original at the same time. The more one looks at how electronic configurations developed the more one is struck by the gradual, piecemeal, and at times almost random development of ideas by various individuals, many of whose names are completely unknown today even among experts. Such a gradualist view flies in the face of the Kuhnian notion whereby science develops in a revolutionary fashion. It goes against the view, fostered in most accounts, of a few heroic characters who should be given most of the credit, like G. N. Lewis, Bohr, or Pauli in the area that is being considered here.

For example, if one looks at the work of the mathematical physicist John Nicholson, one finds the idea of the quantization of electron angular momentum that Bohr seized upon and made his own in his 1913 model of the atom and his electronic configurations of atoms. In looking at the work of the English chemist Charles Bury, one finds the first realization that electron shells do not always fill sequentially. Starting with potassium and calcium, a new shell is initiated before a previous one is completely filled, an idea that is especially important for understanding the chemistry of the fourth period starting with potassium.

Another case is Edmund Stoner, the Cambridge University graduate student who was the first to use the third quantum number to explain that electron shells were not as evenly populated as Bohr had first believed. Instead of a second shell consisting of two groups of four electrons that Bohr favored, Stoner proposed two groups of two and six electrons, respectively. Meanwhile, the Birmingham University chemist Main Smith independently published this conclusion at about the same time. But who has ever heard of Main Smith? The steps taken by these almost completely unknown scientists, Stoner and Main Smith, catalyzed the work of Wolfgang Pauli when he proposed a fourth quantum number and his famous Exclusion Principle.

What I am groping toward, more broadly, is a view of an organic development of science as a body-scientific that is oblivious of who did what when. Everybody contributes to the gradual evolution of science in this view. Nobody can even be said to be right or wrong. I take the evolutionary metaphor quite literally. Just as organic evolution has no purpose, so I believe is true of science. Just as the evolution of any particular biological variation cannot be said to be right or wrong, so I believe to be true of scientific ideas, such as whether electron shells are evenly populated or not. If the idea is suited to the extant scientific milieu, it survives and leads others to capitalize on any aspect of the idea that might turn out to be useful.

Contrary to our most cherished views—that science is the product of brilliant intellects and that logic and rationality are everything—I propose a more prosaic view of a great deal of stumbling around in the dark and sheer trial and error. Of course such a view will not be popular with analytical philosophers, in particular, who still cling to the idea that the analysis of logic and language holds the key to understanding the nature of science. I see it more akin to a craft-like activity inching forward one small step at a time. Language and logic do play a huge role in science but in what I take to be a literally superficial sense. What I mean to say is that language and logic enter the picture at a stage somewhat removed from the occurrence of the initial creative impulse. The real urge to innovate scientifically comes from deeper parts of the psyche, while logic and rational thought only appear at a later stage to tidy things up, or so I will argue.

In recent years many scholars who write about science have accepted limitations of language and rationality, but the same authors have generally tended to concentrate instead on the social context of discoveries. This has produced the notorious Science Wars that so polarized the intellectual world at the close of the 20th century.[8] What I am trying to do is to remain focused on the grubby scientific details of concepts like electron arrangements in atoms, while still taking an evolutionary view which tracks what actually takes place in the history of science.

PREAMBLE

The science covered in this book is centered on some early 20th century atomic physics and chemistry and in particular various precursors to the work of Niels Bohr, and in one case to that of Wolfgang Pauli. A great deal of the action is also centered around the concept of electronic configurations of atoms and their relationship to the periodic table of the elements.

I have long had an interest in minor historical figures and have attempted to highlight their work. I frequently have an uneasy feeling when I read about any particular scientific development which appears as though it came out of nowhere in particular. I always want to know who the precursors might have been and I am invariably driven by the historical trajectory of whatever scientific theory or development I might be studying. This historical "disease" that I also suffer from is sometimes debilitating but, more often than not, proves to be ultimately rewarding in gaining a wider understanding of the science at stake.

As I have admitted before in my previous writings, history was my first academic love while I was still a boy of eight or so years old.[9] The discovery of science only came later when I was around 13 or 14 years old. Another seemingly unrelated interest arrived at the end of my first year as an undergraduate at the University of London. After much soul searching for the meaning of life—no less—I discovered Eastern Philosophy and in particular the Vedanta, Zen, and Taoist philosophies which dealt with the essential unity of everything in nature. A little later a spate of books was published with the aim of connecting Eastern Philosophy to modern scientific discoveries. The best known example was of course Fritjof Capra's *The Tao of Physics*, which has apparently sold many millions of copies and has been through 43 editions and 23 foreign language translations.[10] My own reaction to these proposals was one of initial enthusiasm that soon turned to disappointment and the view that the parallels that were being touted were rather superficial.[11] There is really just one aspect of Eastern philosophy that I am drawn to and it is the appeal to the essential unity of everything in the universe in a way that one seldom encounters in Western Philosophy.

There is one important exception to this statement, namely the philosophy of Baruch Spinoza, to which I was also very much drawn at an early age. According to Spinoza, things in the world that appear to be finite individuals interacting with each other are nothing more than the modifications of a self-caused, infinite substance.[12] Spinoza proposed that everything that exists in the Universe is just one single substance. Spinoza above all other Western philosophers is all about the essential unity of everything. Moreover, the similarities between Spinoza's philosophy and Eastern philosophical traditions have been discussed by many authors starting with the German Sanskrit expert, Theodore Goldstücker, who drew connections between Spinoza's writings and the Vedanta tradition of India.[13]

For many years I maintained an interest in both fields, modern science (along with history and philosophy of science) on the one hand and Eastern Philosophy on the other. My only foray into this area was to write an article entitled *The Tao of Chemistry* since I believe that this field lends itself more to Taoist analogies, including the theme of the co-existence of opposites, than does the field of physics.[14] But I kept these interests strictly separate partly because I feared professional ridicule from colleagues who might accuse me of wanting to grind some ideological axe, or even worse, for fear of being regarded as a New Age pundit.[15]

Having reached the stage of no longer caring much about such perceptions, I have developed a yearning to unify the various intellectual strands in my life. In the present book I attempt to look beyond all the apparent diversity in scientific work and the many contributions by numerous individuals in order to present a unified picture of the underlying forces at play. What I am driving at is a Spinoza-like view of the scientific endeavor, insofar as individual scientists are to be regarded as being essentially one undifferentiated and unified "substance" or entity. I will venture as far as to suggest that the scientific enterprise should be considered as a unified and single organic "entity" with a life of its own in the same way that James Lovelock has spoken of planet earth, or Gaia, as he has christened it.[16] Having said this, I am fully aware of the criticism that Lovelock has endured regarding this view. The most damning of these comes from the likes of biologist Richard Dawkins who points out that Lovelock's Gaia seems to imply a form of teleology, something that is as completely lacking in modern evolutionary theory as it was when Darwin first proposed his theory.[17] Needless to say, I do not intend my own scientific analogue to Gaia to have any sense of teleology.

Although my book is centered on seven scientists I could have added many other lesser-known figures and will in fact be mentioning many more along the way. To be more accurate however, two of the scientists among my chosen seven, van den Broek and Charles Janet, were really amateurs who dabbled in many different disciplines as well as in chemistry and physics. In terms of disciplinary leanings, of the five that may properly be called scientists, two of them, John Nicholson and Edmund Stoner were physicists while the remaining three, Richard Abegg, Charles Bury and John Main Smith were primarily chemists. Finally, in terms of nationalities they consisted of four from Britain and one each from Germany (Abegg), France (Janet) and The Netherlands (van den Broek). Needless to say the predominance of authors who wrote mainly in English reflects my own linguistic limitations and perhaps an unfair bias toward the Anglophone world.[18]

The pivotal person in the case of the five scientists is the physicist Niels Bohr. Each of the five authors either anticipated, corrected, or sought to explore a particular aspect of Bohr's work in atomic physics, even those among them who were clearly chemists. The pivotal person for the remaining two people was Dimitri Mendeleev the leading discoverer of the periodic system. Richard Abegg and Charles Janet attempted to continue the work of Mendeleev in the context of chemical bonding and alternative representations of the periodic system, respectively. All of the seven "scientists" that I will be discussing can be regarded as missing links among the evolutionary branches tracing the developments of various sub-branches of the history of atomic physics and chemistry as it unfolded at the start of the 20th century.

ELECTRONIC CONFIGURATIONS

Another way to regard this book is as a history of a very central concept in modern chemistry and physics, that of the electronic configuration of atoms. As students of chemistry are frequently reminded, if one knows the configuration of any particular atom one has a good way of rationalizing the ways in which it forms bonds as well as being able to rationalize its chemical and physical properties.

The mention of chemical education in turn raises another point that I should mention "up front," namely my own involvement with chemistry teaching and in particular introductory physical chemistry. I have been teaching chemistry for over 35 years, the last 16 of which have been to some of the best students in California at UCLA. I think that this relatively humble task has served to keep my philosophical feet on the ground and has nurtured an interest in how to reach general readers as well as science educators. I hope that science educators may also benefit from reading this book even if they might disagree with some of the more philosophical proposals that I will make.

But in addition to teaching general chemistry I have devoted much research to clarifying the concept of electronic configurations at all levels. For example, I have worked on the vexed question of the occupation and ionization of the 3d and 4s atomic orbitals in the first transition series of metals. This is something of a conundrum that is well known to educators and I believe that I may have contributed to resolving the conceptual issues that accompany attempts to teach this topic in an accurate manner.[19]

Not surprisingly, our understanding of electronic configurations in general has developed in tandem with our knowledge of the structure of the atom. To begin to draw the historical trajectory (but only very briefly at this stage), this journey begins with disagreements as to whether the atom consists of electrons in orbit around the central nucleus (Ernest Rutherford) or whether they might actually be embedded in the atom (J. J. Thomson).

At about this time Niels Bohr, a postdoctoral fellow who spent time with both of these British scientists, developed his model of the atom, in which electrons circle the nucleus in precise orbits that contain specific numbers of electrons. Chemists like Gilbert Lewis and Irving Langmuir, meanwhile, were developing static models in which the electrons were situated in the corners of cubes rather than in circles or spheres. Although their ideas were naïve in terms of the dynamics of the atom, they were able to tackle some issues that the physicists could not.

For example, it was these chemists, rather than Bohr or other physicists, who first ventured to suggest electronic configurations for the transition metals. They were able to do this because they were naturally more familiar with the individual chemical behavior of the atoms of these elements. And this brings us back to the way that the third and fourth shell of electrons around the atom are occupied, a feature that introduces some interesting complications. These complications represented more of a roadblock for physicists than they did for the chemists, although they were eventually able to pick up some hints from the chemists and to provide more refined configurations for the atoms in question.

What I will also do in this book is explore the neglected symbiosis between the fields of chemistry and physics. I think that this neglect occurs for various reasons, the major one being the, to my mind mistaken, view of the extent to which chemistry reduces to physics. This is a view on which I have spent a great deal of time and metaphorical ink in trying to combat. It is the reason why I established the journal *Foundations of Chemistry* that is dedicated to examining the history and philosophy of chemistry as an autonomous and not fully reduced science.[20] There is nothing controversial about the view that physics explores nature at a deeper ontological level. Although this may be true in a trivial sense it does not imply that chemistry is somehow the poor relation of physics or intellectually less prestigious. I could cite a number of famous remarks made by physicists that betray how they sometimes do regard chemistry as indeed being inferior. When Pauli's wife left him, what seems to have worried him the most was that she had left him for a chemist, of all things. A little earlier Rutherford had famously declared that physics was the only true science while comparing chemistry to stamp collecting by which he intended to dismiss it as "mere classification." As a third example, Dirac famously stated that the whole of chemistry had been reduced to quantum mechanics.

Admittedly there is little symbiosis to be found between physics and chemistry these days.[21] The energies that physicists probe are many, many orders of magnitude higher than anything that is of interest to chemists. But about one hundred years ago there was a genuine interaction between the fields. Physicists as well as chemists were dealing with the same energy domain and the same entities, such as atoms and electrons. The goal of explaining the periodic table of the chemical elements provided a testing ground on which physicists could hone their new theories. This was especially true of the likes of J. J. Thomson, Niels Bohr, and Wolfgang Pauli. But the close interaction between chemistry and physics during this period has seldom but been explored in a neutral fashion that does not presume from the outset that physics rules the roost. In fact, chemists, or individuals who adopted a more global chemical perspective, made many key developments as we will see in the cases of van den Broek, Janet, and Main Smith, for example.

THE LITTLE PEOPLE

Within this overriding arc, as I am describing it, there were many contributions from the "little people" who will be the main focus of the book. When Bohr developed his quantum theory of the atom he was building on a hint from the relatively unknown John Nicholson, who first proposed the quantization of the angular momentum of electrons. When Bohr later claimed to settle some difficult configurations such as that of element 72, or hafnium, he was bolstered by some previously published work by the chemist Bury, who had already settled the issue based on empirical evidence.

An earlier episode was Henry Moseley's discovery, or rather his experimental justification, for the concept of atomic number. What is not so well known is that Moseley actually set out to verify the hypothesis of an amateur scientist, namely van den Broek, who was making his living as an economist.

Electronic configurations as we know them now, in terms of four quantum numbers to each electron in an atom, were finally clarified when Wolfgang Pauli announced his exclusion principle. But Pauli was building on a key idea published by the unknown Cambridge University graduate student Edmund Stoner.

One more word about electronic configurations is needed at this point. As mentioned earlier, Lewis's static view of electrons was replaced by various dynamical models of electrons circulating around the nucleus à la Rutherford and Bohr. In an analogous fashion, the notion of a specific electronic configuration is also somewhat static when seen from the perspective of modern chemistry and physics but it still retains an immense usefulness as a first order approximation. Electronic configurations are the starting point for so many explanations, among them the very structure of the periodic table of the elements.

Let me briefly hint at what I mean by this further point. Pauli's assignment of four quantum numbers to each electron and his exclusion principle took place at the tail end of the old quantum theory. During the years, Werner Heisenberg and Erwin Schrödinger independently discovered two forms of a more fundamental quantum mechanics that were soon shown to be equivalent. In addition, attempts to calculate the energy of any particular atom using Shrödinger's approach led Douglas Hartree to develop a method of approximation that now bears his name. When this method was made consistent with the principle that electrons are indistinguishable, by Vladimir Fock, a quite new and more abstract picture of electronic arrangements emerged.[22]

It now became clear that rather than assuming that a particular electron existed in a particular orbital, a better way to imagine the situation was that every electron in an atom was in every orbital at the same time. More recent developments have further undermined the rather static view that will form the subject matter of much of this book: that of a particular electron being in one or the other orbital. For example, accurate calculations in quantum mechanics sometimes demand that we think of fractional occupations of any particular orbitals. Another factor is that accurate calculations require that we imagine the atom as being in a superposition of many contributing configurations.[23] These approaches (which go by the names Configuration Interaction and Perturbation Theory) take us even further away from the "classical picture" that the scientific characters described in the present book could have envisaged.

But these simple and picture-able views continue to shape our imagination and still have immense value in science education as well as science research. This is why the story I will be telling here should be of interest to educators as well as philosophers, historians, and practitioners of the hard sciences.

Finally I would like to mention another motivation for this project. As the history of quantum mechanics is usually presented, it appears as a mainly German affair. Of course if we think of Schrödinger and Pauli then national allegiances must be widened a little to encompass Austria, Switzerland—and Denmark in the case of Bohr. Certainly the Frenchman Louis De Broglie is given due coverage as is the Englishman Paul Dirac (who was of partly French origin). Nevertheless, quantum mechanics is generally regarded as a Germanic affair in the wider sense. As I see it there was a great deal of influential work being carried out in the English-speaking world during this period but this is only evident if one drops the emphasis on the heroic approach to the history of science. It so happens that several of the authors that I will cite wrote primarily in English—and in the case of Janet, in French. Let this serve as a small antidote to the traditional historical account of the pre–World War II history of physics that unfolded roughly between the 1910s and 1930s.

The book goes some way to offering a reconciliation between the internal approach in philosophy of science and the sociological approach, which has been on the rise and which has been blamed for the onset of the Science Wars. My rapprochement with the internal camp comes in the form of paying close attention to the scientific details. I do not however share their enthusiasm for the emphasis on logic or the analysis of the role of language in attempting to understand the nature of science and how it progresses. Similarly my connection with the sociological approach comes from my belief that science progresses as one social entity, one that I claim is a living and evolving organism in the same vein as Lovelock's Gaia view of the earth.

Let me also say that Thomas Kuhn, about whom I will be saying a good deal in my final chapter, seems to have arrived at similar conclusions to mine at least in some respects. As the Kuhn expert Paul Hoyningen-Huene explains,[24]

> Kuhn uses two basic assumptions that import sociology into the philosophy of science. The first assumption states, as I have already said earlier, that communities and not individuals should be seen as the basic agents of science. The second assumption is built upon the first one. —It states that these communities must be characterized by the specific values to which they are committed ....The opposite positions to these assumptions are, of course, logical positivism and critical rationalism. In both these positions, the principal agent, the subject of science, is the individual.
>
> The reason is that in the Kuhnian framework the principal agent in science, its subject, is not the individual but the group. Therefore it seems to me the question of the rationality of theory choice must be asked with respect to groups, not with respect to individuals.[25]

However, I do not take the customary approach of sociologists of science, which generally consists in analyzing the social context of scientific discoveries. Although I believe that social context makes an important contribution to the development of science I am more interested in a radical or literal form of sociology, which regards science as a group activity rather than the work of particular individuals. Or if one must speak of individuals, as I suppose one must, I prefer to include the lesser-known individuals who provided the missing links for the heroic personalities. It is they who provide the glue and continuity that makes the body-scientific more evident than it might otherwise seem.

NOTES

1. My advisor was the noted theoretical chemist, David Buckingham.
2. http://link.springer.com/journal/volumesAndIssues/10698
3. I was fortunate enough to have a three-hour personal meeting with Karl Popper at his home. I wrote briefly about this memorable event in E.R. Scerri, entry on hafnium in *Chemical & Engineering News*, 80th Anniversary Issue, http://pubs.acs.org/cen/80th/elements.html
4. E.R. Scerri, *The Periodic Table, Its Story and Its Significance*, Oxford University Press, New York, 2007.
5. T.S. Kuhn, *The Structure of Scientific Revolutions*, University of Chicago Press, Chicago, 1962, p 169–172.
6. The first such discovery is described in M. Mayor, D. Queloz, "A Jupiter-mass Companion to a Solar-Type Star," *Nature* 378 (6555), 355–359, 1995.
7. This opening passage first appeared as a blog on the Oxford University website, http://blog.oup.com/2015/03/new-philosophy-science-chemistry/
8. Parsons, Keith (ed.), *The Science Wars: Debating Scientific Knowledge and Technology*, Prometheus Books, Amherst, NY, 2003.
9. E.R. Scerri, *Collected Papers on the Philosophy of Chemistry*, Imperial College Press, London, 2008.
10. F. Capra, *The Tao of Physics*, Fontana, London, 1976; G. Zukav, *The Dancing Wu-Li Masters*, Fontana/Collins, London, 1979; M. Talbot, *Mysticism and the New Physics*, Routledge, London, 1981; R. Toben, *Space-Time and Beyond*, Dutton, New York, 1975.
11. Eric Scerri, "Eastern Mysticism and the Alleged Parallels with Physics," *American Journal of Physics*, 57, 687–692, 1989.
12. S. Hampshire, *Spinoza*, Faber and Faber, Oxford, 1951. This book remains the definitive treatment of Spinoza's philosophy.
13. T. Goldstücker, in *Literary Remains of the Late Professor Theodore Goldstücker*, W.H. Allen, London, 1879.
14. E.R. Scerri, "The Tao of Chemistry," *Journal of Chemical Education*, 63, 106–107, 1986.
15. Ibid. I published this article since I believe that chemistry lends itself better to analogies with the coexistence of opposites than the field of physics does. E.R. Scerri, "The Tao of Chemistry," *Journal of Chemical Education*, 63, 2, 106–107, 1986.
16. J.E. Lovelock, "Gaia As Seen through the Atmosphere," *Atmospheric Environment* (Elsevier) 6, 579–580, 1972; J. E. Lovelock, *Ages of Gaia*, Oxford University Press, Oxford, [1988] 1995.
17. Dawkins critique of Lovelock's Gaia appears in R. Dawkins, *The Extended Phenotype: The Gene as the Unit of Selection*, W.H. Freeman, Oxford, 1982.
18. Having said that, French is my first language and Italian my second.

19. E.R. Scerri, "The Trouble with the Aufbau Principle," *Education in Chemistry*, November, 2013, 24–26, http://www.rsc.org/eic/2013/11/aufbau-electron-configuration

20. *Foundations of Chemistry*, http://link.springer.com/journal/10698

21. Of course there is much overlap between the two fields in such areas as nano-technology and related subdisciplines.

22. A. Szabo, N.S. Ostlund, *Modern Quantum Chemistry*. Dover Publishing, Mineola, NY, 1996.

23. Ibid.

24. P. Hoyningen-Huene, The Interrelations between Philosophy, History and Philosophy of Science in Thomas Kuhn's Theory of Scientific Development, *British Journal for the Philosophy of Science*, 43, 487–501, 1992, p. 492.

25. Ibid., p. 495.

A TALE OF SEVEN SCIENTISTS AND
A NEW PHILOSOPHY OF SCIENCE

Chapter 1

Introduction

I was fortunate enough to be invited to give a plenary lecture to a national meeting of the American Physical Society that was held in Denver, Colorado in March 2013. There I also heard a presentation by the leading historian of physics, John Heilbron, who spoke about the Bohr model of the atom in commemoration of its 100th anniversary.

While reviewing the precursors to Bohr's theory, Heilbron mentioned the little known English physicist, John Nicholson, who seemed to have achieved a remarkable success in the year 1911. Nicholson was able to give a quantitative explanation for 9 out of 11 of the unidentified lines in the spectrum of the nebula in Orion, as well as 14 unidentified lines in the spectrum of the solar corona. In addition he predicted some unknown lines in the spectra of both of these astronomical objects; many of these lines were later discovered.

Nicholson achieved all of this on the basis of what soon turned out to be an incorrect theory. He assumed that the electrons circulating around a central nucleus[1] undergo vibrations in a plane perpendicular to their direction of circulation. He also assumed that the spectral frequencies in question corresponded directly to actual mechanical frequencies of the circulating electrons. In the soon-to-follow Bohr model of the atom both of these notions were abandoned entirely.

I began to think about how it is that supposedly incorrect theories can sometimes produce scientific progress, as in the case of Nicholson.

Was it due to sheer luck or perhaps a fortuitous cancellation of errors, or is there some deeper, more significant feature of science that lurks underneath such cases? Of course there is a sense in which such cases are not rare and isolated, but in fact, represent the rule rather than the exception. As the philosopher of science Imre Lakatos once wrote, "All theories are born refuted." That is to say that all theories are eventually shown to be incorrect and yet they generally produce what we term scientific progress when judged from the standpoint of what was known at the time. It has to be admitted that this is a rather odd state of affairs while at the same time seeming to be perfectly commonplace.[2]

I believe that traditional analyses of science tend to elevate the role of logic, rationality, and consistency of theories to excessive levels. I want to urge a more tacit form of development of science in which these aspects are still present but somewhat overshadowed by an underlying thrust that is of an essentially biological origin. The persistent emphasis on the role of logic, for example, is a little surprising given the alleged move away from the logical analysis of theories that was the dominant way of doing philosophy of science in the middle of the 20th century. Modern philosophers of science typically make disparaging remarks about the logical positivist school of philosophy and claim that their own approaches amount to a demotion of the former central role of logic.

This is especially true for those among them who take a thoroughly historical or sociological approach to the study of science. However, by and large, those that continue to identify themselves primarily as philosophers of science do not seem to have given up a belief that logic and rationality are the central governing principles. The approach I present in this book is more radical and at the same time more pedestrian. I claim that science proceeds by almost imperceptible small steps in an evolutionary fashion, not so much through the genius and brilliance of individual scientists but more by a process of trial and error, chance and sheer stumbling around. Above all, I claim that science is a collective enterprise, but not consciously so. Although Big Science has taken an increasingly important role, especially since the middle of the second half of the 20th century, I want to focus on a less consciously collective shared aspect of science. I am referring to science as a collective enterprise in an unwitting fashion in which many individuals, some significant, others far less so, make contributions which are taken up by countless other scientists in the shared growth of the store of scientific knowledge.

As already mentioned I regard science almost as a living organism. The individual contributors often fight and bicker among themselves and this contributes positively to the overall growth and refinement of science. This process, I suggest, is analogous to the struggle among different biological species for survival. But it's more than an analogy, since science is conducted by the human species, which itself is subject to biological evolution at many different levels, be it physiological or mental. Just as the development of the human species is governed by evolutionary forces so too, I claim, is one of the most advanced activities that the species can engage in, namely the doing of science.

MARGINAL AND INTERMEDIATE FIGURES IN THE HISTORY OF SCIENCE

As I have indicated, I am interested in intermediate and lesser-known figures in the history of science such as Anton van den Broek, the pioneer of the concept of atomic number, and Edmund Stoner, who was the first to apply the third quantum number to understanding the physics of the atom and atomic spectra. I believe that John Nicholson provides another good example of such lesser-known, and yet pivotal, figures whose work may be worth studying further. I decided that this would make a good topic for a book to follow up my *A Tale of Seven Elements* in which I examined several scientific controversies among the discoverers of seven elements in the 20th century.[3]

As far as I am aware there has not been any book or study conducted specifically on intermediate and little-known scientific figures working within a particular field. One of the aims of the present book is to fill this gap. The intermediate figures I will focus on are drawn from the fields of physics and chemistry. No doubt other examples could be found from other scientific disciplines but I leave such work to those that have greater expertise than myself in fields other than the physical sciences.

Some readers might object that my examples do not adhere to any particular typology of intermediate figures in science. Nicholson, my main protagonist, was a successful and well-established mathematical physicist at first Cambridge and then King's College, London. Stoner, a research student in Cambridge when he made his important contribution, provides a quite different case. Van den Broek was a lawyer and amateur scientist who was equally well versed in chemistry and physics. As I will argue, the originality of his contribution lay in using chemical criteria having to do with the periodic system rather than purely physical arguments. Bury, another chemist, was based at a provincial university in Wales and made contributions that few chemists have ever heard of.

I believe that these differences are not so important to my project, as I will be explaining in due course. They are all similar in one important respect, namely in contributing to the overall development of different branches of physical science and in being largely forgotten by all but a few historians of science. Their importance, from my point of view, lies in the fact that their contributions stimulated the work of others and that they made up important parts of the body-scientific. The individuals that I have named are the missing links in the evolution of modern atomic theory.

MOVING BEYOND "RIGHT" AND "WRONG" IN SCIENCE

I want to consider the idea of looking beyond theories as being either "right" or "wrong." I will try to examine the development of science as far as possible from far above the contributions of individual theories and individual persons. I will suggest that scientific progress can be regarded as something of a unified giant organism that is constantly evolving and in so doing is experimenting with slightly new ideas and theories. I propose that this may be similar to the way that evolving biological organisms are constantly "trying out" new biological variations and letting nature decide which of them is favorable. One can think of biological evolution as proceeding by conducting exploratory experiments via random mutations that may or may not turn out to be beneficial for any particular living organism. If scientific progress does indeed take place in such an analogous fashion, it may be more appropriate to think of this as happening gradually rather than as a series of abrupt revolutions as Kuhn has argued in his hugely influential book, *The Structure of Scientific Revolutions.*

According to my view, minor figures such as Nicholson are just as important as the supposedly major ones, such as Bohr in that particular case. After all, it is widely acknowledged that Bohr may have taken the crucial notion of the quantization of angular momentum from Nicholson. And even if Nicholson *was* simply wrong about everything he published, he still provided a foil for Bohr, who referred to him frequently, especially in his famous paper announcing the Bohr model of the atom.[4] The historian of science Russell McCormmach, who has produced what is perhaps the only in-depth study of the work of Nicholson, writes,

It must strike many readers as odd on first looking at Niels Bohr's famous 1913 series of articles on the quantum theory of atoms and molecules that a name as unfamiliar as Nicholson turns up so frequently.[5]

As McCormmach notes, the purpose of his study is to show that Nicholson's theory was

> very probably a motivating influence on the direction of Bohr's own developing notions of atomic structure. That Bohr and other contemporaries took Nicholson's work seriously is itself an assurance of Nicholson's historical significance. Nicholson's atomic theory, in a large measure through its probable involvement with the early stages of Bohr's work, played a notable role in the revolution in physical science of the first quarter of the twentieth century.[6]

Not everybody agrees with this view however, as will become clear in due course. This book will also propose another radical thesis, namely a highly impersonal view of scientific development in which personalities, egos, and who might be "right" or "wrong" all become quite irrelevant. All that matters, I suggest, is that the scientific community (or body-scientific, to maintain my biological analogy) should make overall progress. Of course priority and individual success will continue to be valued in science but that is just so that individuals might keep on striving and working hard. In the proposed view, the scientific community consists of a collection of fiercely competing individuals who also make up a seamless and unified single organism.

As mentioned, I draw an analogy to James Lovelock's Gaia hypothesis whereby the earth is regarded as one enormous living organism that, under normal conditions, is able to regulate its own nature—such as the composition of the atmosphere for example. Perhaps I can call my own hypothesis by the name of "SciGaia." But in my case the emphasis will not be so much on the self-regulating aspect as in the essential unity and organic nature of scientific development.

What would it mean if this radical proposal were correct? Consider first the effect of the discoveries made by Copernicus and Galileo and how they have shown us that we are not located at the center of the universe. It is well known that later astronomical discoveries have made our role in the universe less and less significant. Darwin's theory of evolution has demonstrated that our species is not as special as we once may have believed. Even more recently quantum mechanics has revealed that our classical thinking about everyday objects becomes quite irrelevant when it comes to understanding the behavior of matter and radiation at the most fundamental level. As a final example, Kuhn is among the historians and philosophers of science who showed that science does not provide a cumulative march toward the "truth."

Can we bear to go even further and now accept that our heroic individuals are not the basic unit of scientific progress? And I am not just talking about the increasing influence of Big Science such as the enormous team of researchers that recently detected the long-awaited Higgs particle. I am alluding to the deeper idea that it is not only the heroes of science who are responsible for progress but that everybody, including the lesser figures involved in any scientific development, play a fundamentally equal role. In this view all participants are integral parts of one underlying whole and so it makes little sense to distinguish them in the first place. Rather than the named individuals from the pantheon of the history of science, we should focus on the faceless "organism" that we call the scientific community and how it achieves progress in a gradual manner. Here is where the Nicholsons, van den Broeks, and Stoners come into the picture as equal players. And by calling this entity SciGaia I hope to emphasize the essential unity and living organic nature of scientific development that is rather different from the prevailing syntactic or semantic approaches to scientific theories.

The purpose of this book is radical in yet another respect. I will argue that two leading 20th-century philosophers of science have done harm to the history of science by emphasizing discontinuity and swift revolutions. Popper's view of the progress of science via refutations is consistent with radical breaks rather than continuity among successive theories.[7] Kuhn is even more explicit in championing the role of crises in science followed by revolutions, which he likens to political revolutions or gestalt switches in psychology.[8]

But as Kuhn himself also concedes, the more one looks at the details of scientific episodes the more one sees precursors, near misses, half-digested premonitions, and so on. The view of science that I support is an organic one in which scientific knowledge is viewed as one interconnected organism, a living Gaia-like creature possessing many tentacles, branches, and sub-branches. In this view there are no winners or losers in the race to arrive at a better description of Nature. And there are no abrupt scientific revolutions.

There are just hundreds and thousands of worker-bee–like scientists who all contribute to the overall progress. Ultimately there are not even any outstanding personalities, since discoveries can seldom be credited to any particular individual. Of course this view from Mount Olympus, this God's-eye point of view, would seem to be too abstract and too featureless to be of much value. We must therefore come down from Mount Olympus and somehow still examine matters in terms of named individuals who have made contributions to specific scientific episodes, but while still trying to maintain the essentially inter-connected nature of the enterprise.

Some of the cases I propose to consider involve the development of quantum theory in the realm of atomic physics. I concede that the acceptance of quantum theory may look very much like a sharp revision of our basic views and it is of course an example that is frequently cited as the clear-cut case of a scientific revolution.[9] I'm going to examine the work of a few physicists and chemists who were usually considered marginal by science historians as well as some who are considered to have been simply mistaken in their views. I will argue that these contributions severely dampen any Kuhn-like revolutionary claims.

Niels Bohr is generally believed to have been the first physicist to bring the quantum into the study of the atom. Historical accounts already temper this claim by mentioning such names as Nicholson's but quickly add that he and other precursors turned out to be wrong. Although historians of science have long recognized the need to avoid Whiggism, or viewing matters from the perspective of modern-day knowledge, I don't think that even the best of them have gone quite far enough. I believe that a closer examination of cases such as Nicholson's work will serve to support a more radical point of view.

POSSIBLE OBJECTIONS

There are of course many possible objections to my project that I would like to start addressing right from the outset. First, I can almost hear professional historians of science complaining that they are not in fact in the habit of declaring a particular scientist to have been either right or wrong or of apportioning credit. My response to this charge is that I am not just addressing myself to historians but also to scientists at large, to textbook authors, and to the people who present science to a wider public in various ways. In these wider arenas I believe there is frequently more judgment about the worthiness of a particular scientist or claims as to the relative merit of particular individuals. And even within professional history and philosophy of science there are countless debates as to what episode really constituted one particular scientific revolution or another. None of this matters in the more holistic and more organic view that I am suggesting.

While developing my new approach I have frequently wondered how it compares with that of Thomas Kuhn. Looked at in some respects the later Kuhn had already hinted at much of what I wish to say. Is it just a case of his not having gone far enough—or rather of not putting his case in a sufficiently organic manner, for want of a better term? Perhaps so but the question demands closer scrutiny.

On the other hand I see profound disagreement with Kuhn in view of his insistence on discontinuity and revolutionary breaks in the development of science, which seems to contradict any suggestion of an organic evolution of scientific knowledge. Biological evolution is seldom discontinuous, barring catastrophic events such as the extinction of certain species like dinosaurs.

NOTES

1. Nicholson arrived at this idea independently of Rutherford.
2. Since starting to think about this book I have become aware of a thriving subdiscipline in the philosophy of science that addresses some of these questions. I will be commenting on this work in the final chapter of this book.
3. E.R. Scerri, *A Tale of Seven Elements*, Oxford University Press, New York, 2013.
4. N. Bohr, "On the constitution of atoms and molecules [in three parts]. *Philosophical Magazine*, Series 6, 26, 1–25 [I], 476–502 [II], and 854–875 [III].
5. R. McCormmach, "The Atomic Theory of John William Nicholson, *Archives for History of Exact Sciences*, 3, 160–184, 1966, p. 160.
6. Ibid, p. 160.
7. K.R. Popper, *The Logic of Scientific Discovery*, 2nd ed., Routledge, London, 2002.
8. T.S. Kuhn, *The Structure of Scientific Revolutions*, University of Chicago Press, Chicago, 1962.
9. There have been better examples of revolutions such as the Copernican and Darwinian revolutions that Kuhn of course uses as examples.

Chapter 2

John Nicholson

John Nicholson (fig. 2.1) was born in Darlington in Yorkshire in 1881. He attended Middlesbrough High School and then the University of Manchester, where he studied mathematics and physical sciences. He continued his education by going on to Trinity College, Cambridge where he took the mathematical tripos exams in 1904.[1] Nicholson won a number of prizes at Cambridge including the Isaac Newton Scholar Prize for 1906 and was a Smith Prizeman in 1907, as well as an Adams Prizeman in 1913 and again in 1917. His first position was as lecturer at the Cavendish Laboratory in Cambridge, followed by a similar position at Queen's University in Belfast. In 1912 Nicholson was appointed professor of mathematics at King's College, London where he carried out most of his important work. In 1921 he was named fellow and director of studies at Balliol College, Oxford before retiring in 1930 due a recurring problem with alcoholism.

Nicholson was a fellow of various scientific societies, including the Royal Astronomical Society and the Royal Society. In addition he was vice-president of the London Physical Society and president of the Röntgen Society. He died in Oxford in 1955.

THE WORK

Nicholson proposed a planetary model of the atom in 1911 that had certain features in common with those of Jean Perrin, Hantaro

Figure 2.1. John Nicholson.

Nagaoka, and Ernest Rutherford.[2] The chief similarity was that he placed the nucleus at the center of the atom—but it must be emphasized that Nicholson arrived at this conclusion independently of Rutherford and the other physicists just mentioned.

Moreover, the spirit of Nicholson's model had more in common with Thomson's model, which regarded the electrons as being embedded in the positive charge that filled the whole of the volume of the atom. Thomson's later models envisaged the electrons as circulating in rings but still within the main body of the atom. More specifically, the way in which Nicholson's model resembled those of Thomson lies in the mathematical analysis and the concern for the mechanical stability of the system being envisaged.

Where Nicholson's model differed from all previous ones, whether planetary or not, was in his emphasis of astronomical data. Nicholson postulated a series of proto-atoms, as he called them, that would combine to form the familiar terrestrial elements. He believed that the proto-atoms, and the corresponding proto-elements, existed only in the stellar regions and not on the earth. In this thinking Nicholson was part of a British tradition that included William Crookes and Norman Lockyer, each of whom believed in the evolution of the terrestrial elements from matter present in the solar corona and in astronomical nebulae. And like Crookes and Lockyer, Nicholson was an early proponent of the study of spectra for gaining a deeper understanding of the physics of stars as well as the nature of terrestrial elements.

The particular details of Nicholson's proto-atoms were entirely original to him and are represented in the form of a table (fig. 2.2). The first feature to notice is a conspicuous absence of any one-electron atom.[3]

element	symbol	nuclear charge	atomic weight
coronium	Cn	2e	0.51282
'hydrogen'*	H	3e	1.008
nebulium	Nu	4e	1.6281
Protofluorine	Pf	5e	2.3615

Figure 2.2. Nicholson's proto-elements. *Hydrogen was not intended to represent terrestrial hydrogen. Nicholson believed that the latter was a composite of several proto-hydrogen atoms, although he did not specify how many.[4]

This is because Nicholson believed that such a system would be unstable according to an electromagnetic analysis that owed much to the work of Thomson and Larmor.

For Nicholson, the identity of any particular atom was governed by the number of positive charges in the nucleus regardless of the particular number of orbiting electrons present in the atom. Nicholson may thus be said to have anticipated the notion of atomic number that was later elaborated by van den Broek and Moseley. As already mentioned, he argued that a one-electron system could not be stable since he believed this would produce a resultant acceleration toward the nucleus. Little did he know what Bohr would soon do with a one-electron atom. According to Nicholson the two or more electrons adopted equidistant positions along a ring so that the vector sum of the central accelerations of the orbiting electrons was zero. The smallest atom therefore had to have at least two electrons in a single ring around a doubly positive nucleus.

By appealing to his proto-atoms, Nicholson set himself the enormous task of calculating the atomic weights of all the elements and the further task of explaining the unidentified spectral lines in some astronomical objects such as the Orion nebula and the solar corona. It is one of the distinctive features of Nicholson's work that his interests ranged across physics, chemistry, and astrophysics and that he placed great emphasis on astrophysical data above all other data forms.

ACCOUNTING FOR ATOMIC WEIGHTS
OF THE ELEMENTS

Of the four proto-atoms that Nicholson originally considered, he believed that the first of them, coronium, did not occur terrestrially. He therefore set out to quantitatively accommodate the atomic weights of all the elements in terms of just his three remaining proto-atoms, namely his very special hydrogen, nebulium, and proto-fluorine. Before seeing how he carried out this task it is important to consider the relative weights that Nicholson attributed to the proto-atoms. And even before reaching this step it becomes necessary to delve a little further into Nicholson's theory.

Although Rutherford's planetary model had recently been proposed and although it seems to bear the greatest superficial similarity with Nicholson's own planetary model, Nicholson's work is in fact much more indebted to the earlier Thomson model. As is well known to chemists, Thomson regarded the atom as consisting of a diffuse positive charge in which the electrons were embedded as "plums in a pudding."[5] In a later development the electrons were regarded as circulating in concentric rings but still within the main body of the positive charge. According to Thomson the orbital radius of any electron had to be less than the size of the atom as a whole. Nicholson rejected this notion for reasons that were quite independent of the arguments that were being published by Rutherford at about the same time. Nevertheless there is a sense in which Nicholson's atom can be said to have been intermediate between that of Thomson and the later one due to Rutherford. Nicholson retained much of the mathematical apparatus that Thomson had used to argue for the mechanical stability of the atom but demanded that the positive nucleus should shrink down to a size much smaller than the radius of the electrons. The consequence of this move was that Nicholson could no longer use estimates of the size of the atom to fix the radius of the electron orbits. On the other hand, unlike Thomson's atom, Nicholson could use his own atom to give what seems to have been an excellent accommodation of the atomic weights of all the elements and some astronomical spectral lines as will be discussed further below.

We can begin to see precisely how Nicholson's atom was spelled out by considering his expression for the mass of an atom, which he had already published between 1910 and 1911 in a series of articles[6] on a theory of electrons in metals.[7] The expression is

$$m = 2/3(e^2/ar^2)$$

in which m is the mass of an atom, e the charge on the nucleus, r the radius of the electron's orbit, and c the velocity of light. This expression can be simplified to read

$$m \propto e^2/r \qquad \text{(i)}$$

given the constancy of the velocity of light. Nicholson further assumed that the positive charge for any particular nucleus with n electrons would be, ne.

$$ne \propto V$$

Substituting e = ne into (i) $m \propto n^2 e^2/r$ (ii)

He then assumed that the positive charge would be uniformly distributed throughout a sphere of volume V so that

$$ne \propto V$$
or $ne \propto r^3$ (since $V \propto r^3$),
and so $r \propto n^{1/3}$

Substituting into (ii) the nuclear mass would take the form

$$m \propto n^2 / n^{1/3}$$

$$\text{or} \quad m \propto n^{5/3}$$

At this point Nicholson assigned the mass of 1.008 to his proto-atom of hydrogen,[8] which immediately allowed him to estimate the relative masses of the other proto-atoms as follows (fig. 2.3),

Coronium	Cn	=	0.51282
Hydrogen	H	=	1.008
Neptunium	Nu	=	1.6281
Proto-fluorine	Pf	=	2.3615

Figure 2.3. Relative weights of Nicholson's proto-atoms.

From here Nicholson just combined different numbers of these three particles (omitting Cn) to try to obtain the weights of the known terrestrial elements (fig. 2.4). Success apparently came to him from the very start of this approach since he found that the weight of terrestrial helium could be expressed as,

$$He = Nu + Pf = 3.9896$$

a value that compares very well with the weight of helium that was known at the time, namely 3.99.[9]

Nor were Nicholson's calculations of atomic weights confined to the first few elements as shown in the figure 2.3. He was able to extend his accommodation of atomic weights of all the elements up to and including the heaviest known at the time, namely uranium, and to a very high degree of accuracy. For example, figure 2.4 shows his calculations as well as the observed atomic weights for the noble gases. Meanwhile figure 2.5 shows the calculated and observed weights for

Gas	Formula	Calculated atomic weight	Observed atomic weight
helium	Nu+Pf	3.99	3.99
neon	$6(Pf+H)$	20.21	20.21
argon	$5\,He_2$	39.88	39.88
krypton	$5\{Nu_4(Pf+H)_3\}$	83.0	82.9
xenon	$5\{He_4(Pf+H)_3\}$	130.29	130.2

Figure 2.4. Slightly modified table based on a report of Nicholson's presentation. *Nature*, 87, 2189, 501–501, 1911.

H	H	1.008	1.008
He	Nu+Pf	3.99	3.99
Li*	$3Nu + 2H$	6.90	6.94
Be	$3Pf + 2H$	9.097	9.10
B	$2He + 3H$	11.00	11.00
C	$2He + 4H$	12.00	12.00
N	$2He + 6H$	14.02	14.01
O	$3He + 4H$	15.996	16.00
F	$3He + 7H$	19.020	19.00
Ne	$6\,(Pf + H)$	20.21	20.21
Na	$4He + 7H$	23.008	23.01

Figure 2.5. Nicholson's composite atoms for the first 12 elements in the periodic table.

the first eleven elements in the periodic table. The second of these figures did not appear in Nicholson's own papers but in a 1911 article in *Nature* magazine as part of a report on the annual conference of the British Association for the Advancement of Science meeting at which Nicholson had presented some of his findings.

A comment made following the publication of this table is worth quoting in order to see how Nicholson's contemporaries reacted to this work:

> The coincidence between the calculated and observed values is great, but the general attitude of those present seemed to be one of judicial pause pending the fuller presentation of the paper, stress being laid on the fact that any true scheme must ultimately give a satisfactory account of the spectra.[10]

Nicholson promptly rose to the challenge and responded by providing just such an account of the spectra of some astronomical bodies in his next publication.

This contribution involved the hypothetical proto-element nebulium, which Nicholson took to have just four electrons orbiting on a single ring around a central positive nucleus with four positive charges. Like his other proto-elements, Nicholson did not believe that this element existed on the earth but only in the nebulae that had long ago been discovered by astronomers, such as the one in the constellation of Orion. Following a series of intricate mathematical arguments, and building on J. J. Thomson's model of the atom, Nicholson found that he could explain many of the lines in the nebular spectrum that had not yet been explained by others who had invoked lines associated with terrestrial hydrogen or helium (fig. 2.8).

Now this feat could be regarded as a numerological trick, given that it is always possible to explain a set of known data points given enough doctoring of any theory. In fact when it was first publicly proposed at a meeting of the British Society for the Advancement of Science the reaction was indeed one of caution. A report appeared in the magazine *Nature* saying,

> Dr. J.W. Nicholson contributed a paper on the atomic structure of elements, with theoretical determinations of their atomic weights, in which an attempt was made to build up all the elementary atoms out of four prolytes containing respectively 2, 3, 4 and 5 electrons in a volume distribution of positive electricity. Representing the prolytes by the symbols Cn (coronium), H (hydrogen), Nu (nebulium), Pf (protofluorine), the accompanying table indicates the deductions of the author with regard to the composition of several elements, allowance being made for the masses of both positive and negative electrons (figure 5).

Moreover, to avoid any such suspicion scientists usually demand that a good theory should also make successful predictions.[11] Amazingly enough, Nicholson's theory was also able to do just that. In addition to providing a quantitative accommodation of many spectral lines that had not previously been identified, Nicholson was able to make some genuine predictions which were confirmed soon afterward.

There is one crucial feature of Nicholson's model that has made most historians and scientific commentators dismissive of him. Nicholson assumed that each spectral frequency could be identified with the frequency of vibration of an electron in the ring of four electrons. Furthermore, he believed that these vibrations took place in a direction that was perpendicular to the direction of circulation of the electrons around the nucleus (fig. 2.6). The model that was eventually developed by Niels Bohr in 1913 differed fundamentally in that spectral frequencies are regarded as resulting from differences between the energies or frequencies of two different levels in the atom. Bohr's spectral frequencies do not correspond directly to any actual orbital frequency that an electron possesses. And it was this new understanding of the relationship between spectra and energy levels that

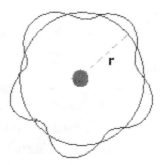

Figure 2.6. Nicholson's atomic model. This figure has been created by the author, although no such diagram was published by Nicholson. As the electron orbits the nucleus Nicholson supposes that it is oscillating in a direction perpendicular to the direction of circulation.

won the day and provided Bohr with one of the main ingredients of his own theory. So on the face of things Nicholson was simply wrong since he based his whole theory on what we now know to be incorrect physics.

But such a view is a typical example of Whiggism and remains at the level of "right" and "wrong" that I am aiming to move beyond. On the view that I am proposing, Nicholson's theory was just one part of the giant organism that we generally call scientific progress, or simply, the scientific community. This organism puts out a new limb, as it were, which turned out not to have any evolutionary advantage. Nevertheless in the context in which it arose there was still a certain degree of progress. Parts of Nicholson's theory do seem to have succeeded, given the many scientists who were impressed by his explanation of the nebular spectrum and his successful prediction of new lines before they had been observed.[12] In addition Nicholson also proposed the notion of quantization of angular momentum which Niels Bohr very soon embraced to much effect.

It is not easy to dismiss Nicholson's accommodation of so many spectral lines and his predictions of some unknown lines. It would not be the first time that progress had been gained on the basis of what later turned out to be an incorrect foundation. Perhaps just enough of Nicholson's overall view was correct enough to allow him to do some useful science. After all, progress cannot be expected to exist uniformly across an entire theory. Typically some parts may be regarded as being progressive while others may lead to degeneration.

And if we take an even wider perspective and consider the *longue durée* in the history of science, surely all scientific progress has been gained on the basis of what later turned out to be incorrect foundations when seen in the light of later scientific views. This is why I claim it may be pointless to ever assert that some particular scientist was right or wrong. What really matters is that science, in the form of the scientific community, makes progress as a whole. Attributing credit to a particular scientist may be important in deciding who should be awarded a prize but does not matter in the overall question of the scientific community gaining a better understanding of the world.

ACCOMMODATING THE SPECTRA OF FOUR NEBULA INCLUDING ORION NEBULA

In this section the manner in which Nicholson was able to assign many unknown lines in the spectrum of the Orion nebula (fig. 2.7) will be examined. First we present a figure containing the spectral lines that had been accounted for in terms of terrestrial hydrogen and helium (fig. 2.8). The dotted lines signify the lines that had not yet been assigned or identified in any way. This situation therefore provided Nicholson with another opportunity to test his theory of proto-atoms and proto-elements.

Figure 2.9 presents unassigned lines in the Orion nebula spectrum.

Figure 2.7. Image of the Orion nebula.

Nebular line	Vacuum tubes	Nebular line	Vacuum tubes
3726.4	...	4101.91	4101.89 Hδ
3729.0	...	4340.62	4340.63 Hγ
3835.8	3835.6 Hη	4363.37	...
3868.88	...	4471.71	helium
3889.14	3889.15 Hζ	4685.73	...
3965.1	3964.9 helium	4740.0	...
3967.65	...	4861.54	4861.50 Hβ
3970.23	3970.25 Hε	4959.05	...
4026.7	...	5006.89	...
4068.8	...		

Figure 2.8. Spectrum of the Orion nebula showing many unassigned lines.

Nebular line	Identification	Nebular line	Identification
3726.4	Nu+	4101.91	Hδ
3729.0	...	4340.62	Hγ
3835.8	Hη, Nu–, Nu++	4363.37	Nu
3868.88	Nu–	4471.71	helium
3889.14	Hζ	4685.73	...
3965.1	helium	4740.0	Nu--
3967.65	Nu++	4861.54	Hβ
3970.23	Hε	4959.05	Nu–
4026.7	Helium?, Nu+	5006.89	Nu
4068.8	Nu–	...	...

Figure 2.9. Nicholson thus accounts for 9 of 11 unidentified lines in figure 2.7.

HOW DID NICHOLSON CALCULATE THE FREQUENCIES OF SPECTRAL LINES?

As in many other features of Nicholson's work his approach was rather simple. He began by assuming that, due to his postulated electron motions, ratios of spectral frequencies correspond to ratios of mechanical frequencies.[13] In mathematical terms he assumed

$$\nu_{\text{spectral line 1}} / \nu_{\text{spectral line 2}} = f_{\text{rotation A}} / f_{\text{rotational B}}$$

where the f values emerged from his calculations, while the ν values were obtained empirically from the spectra in question.

In his 1911 article entitled "The Spectrum of Nebulium," Nicholson also predicted the existence of a new spectral line for the nebulae in question. On page 57 of his article he writes,

> Now the case of k = -2 for the neutral atom has been seen to lead to another line which will probably be very weak. Its wavelength should be 5006.9 × .86939 = 4352.9. It does not appear in Wright's table....

Remarkably enough this prediction was very soon confirmed. In a short note in the same journal in the next year, 1912, Nicholson was able to report that it had been found at a wavelength of 4353.3 Angstroms. The error is just .009% or roughly 1 in 11,110.

> At the meeting of the Society of 1912 March the writer announced the discovery of the new nebular line at λ4353 which had been predicted in his paper on "The Spectrum of Nebulium." A plate of the spectrum of the Orion nebula, on which the line was found, and which had been taken with a long exposure at the Lick Observatory in 1908 by Dr. W.H. Wright, was also exhibited. In the meantime the line has been recorded again by Dr. Max Wolf, of Heidelberg, who has, in a letter, given an account of its discovery, and this brief note gives a record of some of the details of the observation.
>
> The plate on which the line is shown was exposed at Heidelberg between 1912 January 20 and February 28, with an exposure

of 40h 48m. The most northern star of the Trapezium is in the center of the photographed region, and the new line is visible fairly strongly, especially in the spectrum of the star and on both sides.

The wave-length in the Orion nebula, obtained by plotting from an iron curve, is 4353.9, which is of course, too large, as all the lines in this nebula are shifted to greater wave-lengths, on account of the motion of the nebula. But the correction is not so large as a tenth-metre.

The wave-length of the line on the Lick plate, as measured at the Cambridge Observatory by Mr. Stratton, is 4353.3, the value calculated in the paper being 4352.9.[14]

Nicholson experienced a similar triumph over the prediction of a new spectral line which he believed was due to proto-fluorine and which he estimated to have a wavelength of 6374.8 Angstroms. It was soon discovered in the solar corona with a wavelength of 6374.6.

Considered together these successes by Nicholson are indeed rather remarkable. Just to recap, he accounted for 9 of 11 previously unidentified lines in the spectrum of the Orion nebula and 14 of the unidentified spectral lines in the solar corona. In addition, and perhaps more impressively, he predicted two completely unknown lines, one in each of these spectra, and both were promptly discovered and found to have almost exactly the wavelengths that Nicholson had predicted:

Nebulium prediction: 4352.9 A - observation: 4353.3 A - error: 1 in 11,111

Solar corona prediction: 6374.8 A - observation: 6374.6 A - error 1 in 31,745

NICHOLSON'S CALCULATIONS ON THE SPECTRUM OF THE SOLAR CORONA

Nicholson next turned his attention to considering the spectrum of the solar corona (fig. 2.10) that had been much studied and showed numerous lines that had not yet been accounted for (fig. 2.11). In this study Nicholson was even more successful than he had been with the spectrum of the Orion nebula, because he succeeded in accounting quantitatively for as many as 16 unexplained lines.

Figure 2.12 shows the observed frequencies of the lines along with Nicholson's assignments in terms of the atom of proto-fluorine or various ionized forms of the same atom.

NICHOLSON AND PLANCK'S CONSTANT

The manner in which Nicholson arrived at the all-important Planck's constant was by calculating the ratio of the energy of a particle to its

Figure 2.10. Image of the solar corona.

1900	1901	1905	Mean	Intensity
...		5535.8	5535.8	2
...	5304	5303.1	5303.5	20
...	...	5117.7	5117.7	2
5073	...	...	5073	1
4779	...	...	4779	1
4725	...	...	4725	...
4722	...	...	4722	..
4586.3	...	...	4586.3	4
4566.5	4565	...	4566	6
4400	...	...	4400	1
4358.8	...	...	4359	4
4311.3	...	...	4311	2
4230.6	4230.9	4231.1	4231.0	5
4130	...	...	4130	...
...	...	4087.4	4087	...
3987.2	3987	3987.1	3987.1	3
...	3891.2	...	3891	...
3800.8	3801.1	3800.8	3800.9	3
3642.9	...	3642.0	3642.5	2
...	3505	...	3505	...
3461.3	...	...	3461	1
...	3454	...	3454	9
...	3387.9	...	3387.9	12
...	3361	...	3361	...

Figure 2.11. Observed lines in the solar corona at various dates.

frequency and finding that this ratio was equal to a multiple of Planck's constant. Nicholson concluded that Planck's constant therefore had an atomic significance and indicated that angular momentum could only change in discrete amounts when electrons leave or return from an atom. It is worth bearing in mind that up to this point the quantum had only been associated with energy and not with

Coronal line	Suggested origin	Intensity
5535.8	Pf, −2e	2
5073	Pf, −3e (?)	1
4586	Pf, +2e	4
4566	Pf, −e	6
4359	Pf, −2e	4
4311	Pf, +2e	2
4231	Pf, +e	5
4087	Pf, +3e (?)	...
3987.1	neutral atom, +3e (?)	3
3800.9	Pf, +e	3
3642.5	Pf, −e	2
3454	Pf, neutral atom	9
3387.9	Pf, neutral atom	12
3361 (?)	Pf, −3e (?)	...

Figure 2.12. Nicholson's identification of 14 of these lines, using proto-fluorine and ionized of this proto-atom.

angular momentum. Nicholson was in fact the first person to make this association, in what would soon become an integral aspect of Bohr's theory of the hydrogen atom.

In the case of the proto-fluorine atom, Nicholson calculated the ratio of potential energy to frequency to be approximately

$$\text{Potential energy} / \text{frequency} = 154.94 \times 10^{-27} \text{ erg seconds} = 25h$$

In arriving at his result Nicholson had used the measured values of e and m, the charge and mass of the electron. However, his method still did not provide a means of estimating the radius of the electron and was therefore forced to eliminate this quantity from his equations, a problem that he overcame a little later.

Nicholson then proceeded to calculate the ratio of potential energy to frequency in proto-fluorine with one or two fewer electrons and found 22 h and 18 h, respectively. He noted that the three values for Pf, Pf⁺, and Pf²⁺ were members of a harmonic sequence,

$$25, 22, 18, 13, 7, 0.$$

He then divided each value by the number of electrons in the atom to find the Planck units of angular momentum per electron to be

$$5, 5.5, 6, 6.5, \text{ and } 7.$$

This led him to a general formula for the angular momentum of a ring of n electrons as

$$\frac{1}{2}(15-n)n$$

This formula in turn allowed Nicholson to fix the values of the atomic radius in each case and since angular momentum did not change gradually this implied that atomic radius too would be quantized.

Several authors have traced the manner in which Bohr picked up this hint of quantizing angular momentum.[15] This feature was not present in Bohr's initial atomic model and he incorporated it over a series of steps following a close study of Nicholson's papers. Bohr also spent a good deal of time trying to establish the connection between his own and Nicholson's atomic theory.

REACTIONS TO WORK OF NICHOLSON

As mentioned earlier, the historian of physics John Heilbron stated in a recent plenary lecture to the American Physical Society that the success of Nicholson's work on nebulium had been "spectacular." He also commented on how it had served as a motivation for Bohr's work.[16] But looking at the literature in physics and the history of physics one finds a remarkable range in the views expressed about Nicholson's work. The following is a brief survey of these varied reactions.

Initially the commentators tended to praise Nicholson to a large extent. For example, following a meeting held in Australia in 1914, W.M. Hicks remarked,

> Nicholson's calculated frequencies and the observed lines were "so close and so numerous as to leave little doubt of the general correctness of the theory.... Nicholson's theory stands alone as a first satisfactory theory of one type of spectra."[17]

In a paper published at the end of 1913, William Wilson observed that Nicholson had "used the quantum hypothesis with extraordinary success in his valuable investigations of the sun's corona."[18]

Here is how physics historian Abraham Pais saw the relationship between Bohr and Nicholson some time later,

> Bohr was not impressed by Nicholson when he met him in Cambridge in 1911 and much later said that most of Nicholson's work was not very good. Be that as it may, Bohr had taken note of his ideas on angular momentum, at a crucial moment for him.... He also quoted him in his own paper on hydrogen. It is quite probable that Nicholson's work influenced him at that time.[19]

Returning to Heilbron,

> The success of Nicholson's atom bothered Bohr. Both models assumed a nucleus, and both obeyed the quantum; yet Nicholson's radiated—and with unprecedented accuracy—while Bohr's was, so to speak, spectroscopically mute. By Christmas 1912, Bohr had

worked out a compromise: his atoms related to the ground state, when all the allowed energy had been radiated away; Nicholson's dealt with earlier stages in the binding.... Just how a Nicholson atom reached its ground state Bohr never bothered to specify. He aimed merely to establish the compatibility of the two models. The compromise with Nicholson was to leave an important legacy to the definitive form of the theory.[20]

Later in the same paper Bohr proposed other formulations of his quantum rule, including, with full acknowledgement of Nicholson's priority, the quantization of the angular momentum.[21]

Another historian-philosopher of physics, Max Jammer, wrote

It should also be pointed out that Nicholson's anticipations of some of Bohr's conclusions were based, as Rosenfeld has pointed out, on *the most questionable and often even fallacious reasoning.*[22] [My italics].

Now for one last commentator, Leon Rosenfeld, who I think brings out some further interesting aspects. In his introduction to a book by Niels Bohr to celebrate the 50th anniversary of the 1913 theory of the hydrogen atom Rosenfeld writes,

The ratio of the frequencies of the two first modes happens to coincide with that of two lines of the nebular spectra: this is enough for Nicholson to see in this system a model of the neutral "nebulium" atom; and as luck would have it, the frequency of the third mode, which he could then compute, also coincided with that of another nebular line, which—to make things more dramatic—was not known when he made the prediction in his first paper, but was actually found somewhat later.[23]

From the mathematical point of view Nicholson's discussion of the stability conditions for the ring configurations and of their modes of oscillation is an able and painstaking piece of work; but the way in which he tries to apply the model ... must strike one as unfortunate accidents...[24]

In the third paper, however, published in 1912, occurs the first mention of Planck's constant in connection with the angular momentum of the rotating electrons: again here there is no question of any physical argument, but just a further display of numerology.[25]

Bohr did not learn of Nicholson's investigations, as we shall see, before the end of 1912, when he had already given his own ideas of atomic structure their fully developed form.[26]

By contrast [with Nicholson] the thoroughness of Bohr's single-handed attack on the problem and the depth of his conception will appear still more impressive.[27]

There is clearly no "fence sitting" here to give Nicholson any benefit of the doubt. Rosenfeld does not even believe that a cancellation of errors might have given Nicholson his apparent early success. But looking into the life of Rosenfeld a little explains some of this reaction. Rosenfeld was without doubt one of Bohr's leading supporters; he also acted as the spokesperson for Bohr's Copenhagen interpretation of quantum mechanics for the last 30 or so years of Bohr's life. Moreover, Rosenfeld is known to have been an especially vitriolic and harsh critic in spite of his having a shy, retiring personality. His fellow Belgian and once collaborator, the physical chemist Ilya Prigogine, described him as being a "paper tiger."[28] So it is hardly surprising that Rosenfeld championed Bohr against any claims from people that he regarded as imposters or anyone who might try to steal even a little of the thunder from Bohr.

The views of Rosenfeld can be contrasted this with those of Kragh, a contemporary historian and like Bohr a Dane,

> No wonder Bohr, when he came across Nicholson's atomic theory found it to be interesting as well as disturbingly similar to his own ideas. Nicholson's atom was a rival to Bohr's and Nicholson was the chief critic of Bohr's ideas of the quantum atom.[29]

But let us say, for the sake of argument, that Rosenfeld is right and that Nicholson's work was completely worthless. Even if this were true, I contend that Nicholson's publications contributed to Bohr's developing his own atomic theory for the simple reason that Nicholson served as his foil. In some places Bohr is quite dismissive of Nicholson's work, such as when writing to his Swedish colleague Carl Oseen, where he describes Nicholson's work as "pretty crazy" while adding,

> I have also had discussion with Nicholson: He was extremely kind but with him I shall hardly be able to agree about very much.[30]

In other places Bohr shows Nicholson considerably more respect, such as when writing to Rutherford while he was on the point of submitting his famous trilogy paper that was published in 1913.

> It seems therefore to me to be a reasonable hypothesis, to assume that the state of the systems considered in my calculations is to be identified with that of atoms in their permanent (natural) state.... According to the hypothesis in question the states of the system considered by NICHOLSON are, contrary, of a less stable character; they are states passed during the formation of the atoms, and are states in which the energy corresponding to the lines in the spectrum characteristic for the element in question is

radiated out. From this point of view systems of a state as that considered by Nicholson are only present in sensible amount in places in which atoms are continually broken up and formed again; i.e. in places such as excited vacuum tubes or stellar nebulae.[31]

In another passage from a letter to his brother Harald, Niels Bohr writes

> Nicholson's theory is not incompatible with my own. In fact my calculations would be valid for the final chemical state of the atoms, whereas NICHOLSON would deal with the atoms sending out radiation, when the electrons are in the process of losing energy before they have occupied their final positions. The radiation would thus proceed by pulses (which much speaks well for) and NICHOLSON would be considering the atoms while their energy content is still too large that they emit light in the visible spectrum. Later light is emitted in the ultraviolet, until at last the energy which can be radiated away is lost.... (Bohr to Harald ...)[32]

If Bohr was the "winner," Nicholson emerges as the equally necessary "loser." Furthermore, after Bohr had published his three-part article, Nicholson continued to press him in a number of further publications. If we must speak in terms of winners and losers, there are no "winners" without the presence of "losers." But as in all walks of life, it is not just about winning, but more about partaking. There would be no athletic races for spectators to watch if the "losers" were not even to participate in the race. The very terms "winner" and "loser" are necessarily codependent in the scheme of any scientific debate.

Now this picture that I have painted would seem to raise at least one obvious objection. If all competing theories are allowed to bloom because there is no such thing as a right or wrong theory, how would scientists ever know which theories to utilize and which ones to ignore? I think the answer to this question can be found in evolutionary biology. Nature has the means and ways of finding the best way forward. Just as any physical trait with an evolutionary advantage eventually takes precedence, so the most productive theory will eventually be adopted by more and more scientists in a gradual, or perhaps even in a trial and error fashion. The theories that lead to the most progress will be those that garner the largest amount of experimental support and which provide the most satisfactory explanations of the facts. This entire process will not be rendered any the weaker even if one acknowledges an antipersonality and anti–"right or wrong view" of the growth of science.

More generally, I believe that the two aspects can coexist quite happily. Scientists on the ground can, and regularly do, fight things out to establish the superiority of their own views, their claims to priority, and so on. But the march of progress, to use an old-fashioned term, does not care one iota about these human squabbles. And it is the overall arc of progress that really matters, not whose egos are bruised or who obtains the greater number of prizes and accolades.

HOW WAS ANY OF THE SUCCESS POSSIBLE GIVEN THE LIMITATIONS OF NICHOLSON'S THEORY?

Having examined the apparent successes of Nicholson's theory I must still ask how any of this was even possible given what we now know of his ideas. Here is a brief list of what seems to be patently "wrong" with Nicholson's scheme: First of all, the proto-elements like nebulium that he postulated do not exist. Second, he "wrongly" identified mechanical frequencies of electrons with spectral frequencies and moreover wrongly assumed that theses oscillations took place at right angles to direction of electron circulation. Third, Nicholson's electrons were all in one single ring, unlike the subsequent Bohr model in which they are distributed across different rings or shells.

So in the light of modern knowledge, Nicholson was making several false assumptions. It is not as though they were even approximately correct—they were downright false. And yet he achieved remarkable success, at least according to most commentators whose views were quoted.

Are cases such as Nicholson exceptional or can other examples be found in the history of science? If one accepts that all, or most, theories are eventually refuted one has to concede that the progress of science implies that "wrong theories" regularly lead to progress!

CONCLUSIONS

Nicholson is all but forgotten in the history of science, but I don't believe it is because he was wrong in many of his basic assumptions. In spite of holding incorrect assumptions concerning the structure of the atom and so on, Nicholson was still able to make a number of highly successful accommodations of known data and predictions of completely unknown information. He is forgotten because history continues to favor heroes such as Niels Bohr and because the history of science focuses primarily on individual contributions instead of recognizing the truly collective nature of scientific research. Nicholson's idea of the quantization of angular momentum was key to Bohr's subsequent progress in the development of atomic physics. Nicholson was part of the organic manner in which science evolves in general, or in this case, the way that atomic physics evolved. He was an important "missing link" between the old classical physics and the new quantum theory and the way it was applied to the atom.

Needless to say, if Nicholson had not been the first to propose the quantization of angular momentum somebody else would probably have done so. I am not necessarily trying to rehabilitate Nicholson's role, but merely wishing to highlight the crucial and catalytic role that is often played by the "little people" in science. Moreover it is quite conceivable that the history of atomic physics might have taken a different path, perhaps one not involving the quantization of angular momentum. The fact remains that it did, and that Nicholson played an undeniable role in what actually took place. My main point, once again, is the organic and evolutionary way in which science develops. It is only in retrospect that priority is attributed to certain contributors. Given our limitations in attempting to reconstruct what is a highly organic and interconnected growth process, it is hardly surprising that we tend to simplify the story by latching onto the leading players in any particular scientific episode.

NOTES

1. The mathematical tripos was a distinctive written examination of undergraduate students of the University of Cambridge, consisting of a series of examination papers taken over a period of eight days in Nicholson's time. The examinations survive to this day although they have been reformed in various ways. A. Warwick, *Masters of Theory: Cambridge and the Rise of Mathematical Physics*, The University of Chicago Press, Chicago, 2003.
2. E.R. Scerri, *The Periodic Table, Its Story and Its Significance*, Oxford University Press, New York, 2007.
3. Nicholson rejected a one-electron atom because he believed that at least one more electron was needed to balance the central acceleration of a lone electron. See p. 163 of McCormmach, "The Atomic Theory of John William Nicholson" (*Archives for History of Exact Sciences*, 3, 160–184, 1966) for a fuller account.
4. Nicholson's list of proto-elements was extended to include two further members in 1914 when he added proto-hydrogen with a single electron and archonium with six orbiting electrons.
5. It turns out that the name "plum pudding" was never used by Thomson nor any of his contemporaries. A.A. Martinez, *Science Secrets*, University of Pittsburgh Press, Pittsburgh, PA, 2011. Because of the currency of the term I have continued to refer to it as such.

6. Nicholson's theory of metals appears in, J.W. Nicholson, "On the Number of Electrons Concerned in Metallic Conduction," *Philosophical Magazine*, series 6, 22, 245–266, 1911.

7. Interestingly, Niels Bohr's academic career also began in earnest with his development of a theory of electrons in metals.

8. This step seems a little odd given Nicholson's statements to the effect that hydrogen the proto-atom is not necessarily the same as terrestrial hydrogen. In using a mass of 1.008 he surely seems to be equating the two "hydrogens."

9. The error amounts to approximately .3 of one percent. Moreover, Nicholson takes account of the much smaller weight of electrons in his atoms. After making a correction for this effect he revises the weight of helium to 3.9881 (or to three significant figures 3.99) in apparent perfect agreement with the experimental value. Such was the staggering early success of Nicholson's calculations. See J.W. Nicholson, "A Structural Theory of the Chemical Elements," *Philosophical Magazine* series 6, 22, 864–889, 1911, pp. 871–872.

10. Anonymous, *Nature*, October 12, 1911, p. 501.

11. There is nevertheless a long-standing discussion in the philosophy of science regarding the relative worth of temporal predictions as opposed to accommodations, or retro-dictions as they are sometimes termed. See S.G. Brush, *Making 20th Century Science*, Oxford University Press, New York, 2015.

12. As expressed in Heilbron's American Physical Society lecture in Denver CO, March, 2013.

13. The detailed calculations can be found in Nicholson's articles, J.W. Nicholson, "The Spectrum of Nebulium," *Monthly Notices of the Royal Astronomical Society, (London)*, 72, 49–64, 1911; "A Structural Theory of the Chemical Elements," *Philosophical Magazine*, (6), 22, 864–889, 1911; "The Constitution of the Solar Corona I, Protofluorine," *Monthly Notices of the Royal Astronomical Society (London)*, 72, 139–150, 1911; "The Constitution of the Ring Nebula in Lyra," *Monthly Notices of the Royal Astronomical Society (London)*, 72, 176–177, 1912; "The Constitution of the Solar Corona II, Protofluorine," *Monthly Notices of the Royal Astronomical Society, (London)*, 72, 1677–692, 1912; "The Constitution of the Solar Corona III," *Monthly Notices of the Royal Astronomical Society (London)*, 72, 729–739, 1912.

14. J.W. Nicholson, "On the New Nebular Line at λ 4353," *Monthly Notices of the Royal Astronomical Society (London)*, 72, 693–693, 1912.

15. J. Heilbron, T.S. Kuhn, "The Genesis of Bohr's Atom," *Historical Studies in the Physical Sciences*, 1, 211–290, 1969.

16. J. Heilbron's APS plenary lecture, March, 2013.

17. McCormmach, "The Atomic Theory of John William Nicholson," *Archives for History of Exact Sciences*, 3, 160–184, 1966, p. 183.

18. Ibid., p. 184.

19. A. Pais, *Niels Bohr's Times, In Physics, Philosophy, and Polity*, Oxford University Press, Oxford, 1991, p. 145.

20. J. Heilbron, "Lectures in the History of Atomic Physics, 1900–1922." In *History of Twentieth Century Physics*, C. Weiner, ed., 40–108, Academic Press, New York, 1977, p. 69.

21. Ibid., p. 70.

22. M. Jammer, *The Conceptual Development of Quantum Mechanics*, McGraw-Hill, New York, 1966, p.73.

23. L. Rosenfeld, in preface to N. Bohr, *On the Constitution of Atoms and Molecules*, W.E. Benjamin, New York, 1963, p. xii.

24. Ibid., p. xii.

25. Ibid., p. xiii.

26. Ibid.

27. Ibid., pp. xiii–xiv.

28. Prigogine was born in Russia but emigrated to Belgium.

29. H. Kragh, *Niels Bohr and the Quantum Atom: The Bohr Model of Atomic Structure* 1913–1925, Oxford University Press, Oxford, 2012, p. 27.

30. Bohr to Oseen, December 1, 1911, in Rud Nielsen, *Bohr Collected Works*, Vol 1, (ref 9), pp. 426–431.

31. N. Bohr letter to Rutherford, 1913 cited in L. Rosenfeld, preface to N. Bohr, *On the Constitution of Atoms and Molecules*, W.E. Benjamin, New York, 1963, p. xxxvii.

32. In both instances where I have written "my," Bohr had actually written "his"—which must surely be typos.

Chapter 3

Anton Van den Broek

The second intermediate figure that I will consider is rather different from John Nicholson in several respects. I am referring to Anton van der Broek, a Dutchman who made important contributions to atomic physics about 100 years ago (fig. 3.1). Whereas Nicholson was in all respects a professional mathematician-scientist, van den Broek was neither a professional nor an academic in the usual sense of the term, in that he never held a university appointment. His training was in the law and in economics—more specifically, econometrics. As a result van den Broek was rather skilled in handling numerical information, something that gave him an advantage in dealing with some questions in physics, as I will be arguing. Before moving onto his scientific work it is worth noting that van den Broek was an unusually well traveled man, both within his native Holland and in Europe as a whole.

Van den Broek was born in Zoetermeer, Holland in 1870 and graduated from the University of Leyden with a law degree. This period included a three-year interruption during which he also studied law at the Sorbonne in Paris. Among his varied interests were the French prison system and medieval architecture. In 1895 he obtained a doctorate in law, also from the University of Leyden. The following year he married Elisabeth Mauve the daughter of the painter Anton Mauve who was a cousin, by marriage, of Vincent van Gogh.

Van den Broek worked in The Hague as a solicitor until 1900, after which his interests appear to have changed. He began by studying

Figure 3.1. One of the few extant photographs of Anton van den Broek.

mathematical economics in Vienna, with Carl Menger, while also spending time in Leipzig and Berlin. In 1903 he turned to physics, the periodic system of the elements, and the structure of the atom. He published his first article on the periodic table in 1907 while living in Noordwijk in the Netherlands. During his most productive period between 1911 and 1915 almost every article or note that van den Broek published appeared with a different address, none of them university departments.

In spite of the differences between van den Broek and Nicholson, and indeed most of the people in this book, van den Broek shared the characteristic of having made an important, if short lived, contribution to the development of the body of scientific knowledge. Briefly put, he was the first to realize, and to elaborate, the key concept of atomic number, that is a set of integers that characterize each of the known chemical elements, even those that are yet to be discovered.

I believe that a further reason why his contributions have been neglected and even downplayed is that they were couched in an essentially chemical point of view. Van den Broek's approach was to consider the periodic table as a whole and to think about the relationship between all the elements, rather than adopting the perspective of physicists, with their preference for focusing on just one or two substances or some specific property that they might have.

This was not because van den Broek was a chemist, since as mentioned already he was an economist. Nevertheless he appears to have had a passion for the periodic table, which motivated the contributions that we are about to examine. Moreover, the fact that he dwelt in an essentially chemical arena, containing all the elements, seems to have contributed further to his neglect by historians and commentators who have tended to favor physics over chemistry, which is not surprising since physics is regarded as the pinnacle of the reductionist approach. As I will try to show, many commentators are far more willing to credit the discovery of atomic number to physicists like Henry Moseley or Niels Bohr in their apparent eagerness to write van den Broek out of the story altogether.

I believe such a move to be a mistake, somewhat akin to regarding rationality as the "be all and end all" of the scientific method, while relegating experimentation, technique, intuition, guesswork, and serendipity as playing lesser and perhaps even inferior roles.

Let me begin where van den Broek himself did, by considering the periodic table of the elements. Its discovery took place in the 1860s at the hands of at least six scientists in different parts of the world, although the Russian chemist Dimitri Mendeleev almost invariably receives most of the credit for his own periodic table published first in 1869.[1] Indeed this case could be regarded as a good example of multiple discovery as I have already claimed elsewhere.[2]

Van den Broek's first article, published in 1907, consists of an analysis of the periodic table whose originality would eventually pay great dividends in atomic physics as well as chemistry. Another ingredient in the same article consisted of van den Broek's attempt to understand the recent work of perhaps the leading atomic physicist of the period, Ernest, later Count Rutherford, who lived in New Zealand; Cambridge, England; Montreal; and then again in Cambridge, where he directed the Cavendish Laboratory.

Among his many discoveries Rutherford found that some atoms disintegrated to produce α and β particles (as he named them). Rutherford proceeded to study these emanations intensely in the hope of discovering their nature. In 1906 he published an article in which he presented three possibilities for the nature of the α particles. He proposed that they were either

(1) molecules of hydrogen,

(2) helium atoms with twice the electric charge carried by a
hydrogen atom, or

(3) half of a helium atom with a charge of twice the hydrogen
atom.

While Rutherford could confidently rule out option (1) because
he thought it too unstable, he was not prepared to venture a preference between options (2) and (3).

Meanwhile the mere "amateur" van den Broek felt no such con-
straints and, not for the first time, used the work of the "professionals"
to great effect. Van den Broek embraced the third option and wrote,

> Where the experiment fails, only pure speculation remains, and
> so it would seem reasonable to try to see whether this helium
> atom, or the "half helium atom" (let's say free Alphon as a "half"
> atom is an absurdity) would be better suited to act as a primary
> element, than the Prout H-atom ever could.[3]

He then combined this postulated alphon particle with his knowledge of the periodic table in the following elegant manner. Van den
Broek supposed that the atoms of all the elements, culminating in
that of uranium, the heaviest, were made of a particular number of
alphons. Curiously, William Prout's hydrogen atom was the one element that did not sit comfortably in van den Broek's scheme, although
he did not let this detail derail his proposal. Instead he simply omitted hydrogen from his periodic table. This conspicuous absence of
hydrogen is rather inevitable however in van den Broek's scheme.
Having just made a statement about the absurdity of speaking of half
an atom, van den Broek cannot bring himself to equate the hydrogen
atom, which has a weight of one unit, with half of an alphon as he
would need to do for the sake of consistency.

So according to van den Broek the atoms of the elements consist of a series corresponding to the even whole numbers from 2 up to 240 such that there should be a total of 120 elements, each one made up of a whole number of alphons or a particle with a weight of 2 units.

The reason van den Broek even considered this possibility is that in different experiments Rutherford and Charles Barkla had independently concluded that the charge of any atom is approximately half of its atomic weight. Van den Broek just went further in this direction by initially supposing that the relationship was exact and that it applied to all the elements of the periodic table. Here then is an example of his more holistic, and perhaps chemical, way of thinking in considering all elements on an equal footing rather than the more specific approach of the physicists.

At the time van den Broek was writing there were about 80 known elements. So how was he proposing to populate the remaining 40 or so spots in his new periodic table? Fortunately this did not present

TABLE 1

	VII	0	I	II	III	IV	V	VI
1	2* (α)	4 He	6 Li	8 Be	10 B	12 C	14 N	16 O
2	18 F	20 Ne	22 Na	24 Mg	26 Al	28 Si	30 P	32 S
3	34 Cl	36 Ar	38 K	40 Ca	42 Sc	44 Ti	46 V	48 Cr
4	50 Mn	52	54	56 Fe	58 Co	60 Ni	62	64
5	66	68	70 Cu	72 Zn	74 Ga	76 Ge	78 As	80 Se
6	82 Br	84 Kr	86 Rb	88 Sr	90 Y	92 Zr	94 Nb	96 Mo
7	98	100	102	104 Ru	106 Rh	108 Pd	110	112
8	114	116	118 Ag	120 Cd	122 Jn	124 Sn	126 Sb	128 Te
9	130 J	132 Xe	134 Cs	136 Ba	138 La	140 Ce	142 Nd	144 Pr
10	146	148	150 Sa	152	154 Gd	156	158 Tb	160
11	162	164	166 Er	168 Tu	170 Yb	172	174 Ta	176 W
12	178	180	182	184 Os	186 Ir	188 Pt	190	192
13	194	196	198 Au	200 Hg	202 Tl	204 Pb	206 Bi	208
14	210	212	214	216	218	220	222	224
15	226	228	230	232 Ra	234	236 Th	238	240 U

* Theoretical atomic weight.

Figure 3.2. Van den Broek's periodic table of 1907.

too much of a problem, at least in principle, since many new radioactive species were being rapidly discovered, although whether they should be regarded as genuine elements or not was in some doubt. In sum, the article of 1907 does not show any sign of the concept of atomic number, unless one divides each of the atomic weights in van den Broek's table by two to obtain a sequence of values from one to 120 (fig. 3.2). But the Dutchman did not yet take this important step, although he already implied it when using the Rutherford-Barkla formula in arriving at his new periodic table.

THE ARTICLE OF 1911 AND A LETTER TO
NATURE MAGAZINE

In 1911 van den Broek took a new step toward the concept of atomic number. Building upon an obscure passage in an article by Mendeleev, he attempted to design a three-dimensional, or cubic, periodic system (fig. 3.3). As van den Broek described it, his new periodic system was a

> cubic system, consisting of five major periods, each comprising three small periods of 8 elements, and therefore a cube five places high, three places deep and eight places wide, with 120 locations.[4]

In each case, the elements shown diagonally are those that are supposed to be represented along the third dimension. Rather significantly the postulated alphon particle of 1907 is not even mentioned in this article.

All that remains is the all-important idea that successive elements differ from each other by two units of weight as compared with Mendeleev's and many other periodic tables which show varying differences ranging from one to four units, and in some cases even higher intervals.

TABLE 2

		0			I			II			III			IV			V			VI			VII		
		1	2	3	1	2	3	1	2	3	1	2	3	1	2	3	1	2	3	1	2	3	1	2	3
A	1	He			Li			Be			B			C			N			O			F		
A	2		Ne			Na			Mg			Al			Si			P			S			Cl	
A	3			Ar			K			Ca			Sc			Ti			V			Cr			Mn
B	1	Fe			Co			Ni			Cu			−			−			−			Zn		
B	2		−			−			−			Ga			Ge			As			Se			Br	
B	3			Kr			Rb			Sr			Y			Zr			Nb			Mo			Ru
C	1	Rh			Pb			−			−			Ag			−			−			Cd		
C	2		−			−			−			In			Sn			Sb			Te			J	
C	3			Xe			Cs			Ba			La			Ce			Nd			Pr			(Sm)
D	1	(Eu)			(Gd_1)			(Gd_2)			(Gd_3)			(Tb_1)			(Tb_2)			(Dy_1)			(Dy_2)		
D	2		(Dy_3)			(Ho)			(Er)			(Tu_1)			(Tu_2)			(Tu_3)			(Yb)			(Lu)	
D	3			−			−			−			−			−			Ta			W			Os
E	1	Ir			Pt			Au			Hg			Tl			Bi			Pb			−		
E	2		−			−			−			−			−			−			−			−	
E	3			−			−			Ra			−			Th			−			U			−

Figure 3.3. Van den Broek's cubic periodic table of 1911.

In the same year van den Broek published a remarkably short but pregnant statement in *Nature* magazine. It is here that we can begin to see the concept of atomic number taking a definite shape. This is the article that can be said to mark the birth of atomic number, if one insists on making such an identification.

Van den Broek repeats the fact that two independent lines of experimentation, due to Rutherford and Barkla, respectively, have pointed to the simple and approximate relationship between the charge on an atom and its atomic weight, namely,

$$\text{charge} \approx A / 2$$

By reference to his new cubic table and his prediction that a total of 120 elements should exist, van den Broek concludes by saying,

> If this cubic periodic system should prove to be correct, then the number of possible elements is equal to the number of possible

permanent charges of each sign per atom, or to each possible permanent charge (of both signs) per atom belongs a possible element.[5]

Seen from a slightly different perspective, van den Broek is suggesting that since the charge on an atom is half of its atomic weight, and since the weights of successive elements differ by two units in a stepwise fashion, then the charge on an atom defines its position in the periodic table. Neither Rutherford, nor Barkla, nor anybody else it would seem, had concerned themselves with the elements in the periodic table as a whole and consequently they had all missed this key feature.

Said yet another way, Rutherford and Barkla recognized that

$$charge \approx A \, / \, 2$$

while van den Broek went further in seeing that

$$charge \approx A \, / \, 2 = atomic \; number$$

An interesting comment was made a good deal later by the physicist and author Abraham Pais:

> Thus based on an incorrect periodic table and on an incorrect relation ($Z \approx A/2$), did the primacy of Z as an ordering number of the periodic table enter physics for the first time.[6]

Whereas Pais seems to regard this situation as something of an anomaly, I believe it speaks directly to the main thesis of the present book. Two mistakes can make a right—or better still, there are no mistakes but just a groping in the darkness with no particular goal toward a fixed external truth. Furthermore the incremental and evolutionary small steps forward, such as these contributions from van den Broek, can come from the most unlikely of people and yet can still propel science forward.

Indeed van den Broek went further in highlighting and articulating the importance of atomic number, rendering his contribution all the more undeniable. Just as the 1911 article involved the abandonment of the alphon particle, so the next important article (of 1913) involved the abandonment of the cubic periodic table which he had regarded as being so important just a couple of years before. Once again we should not despair at this kind of promiscuous and almost random thinking since it is the rule rather than the exception in the growth of scientific knowledge.

In 1913 the cubic table was replaced by a rather elaborate two-dimensional system, shown in figure 3.4. Van der Broek now also made clearer a statement that for the first time mentions a serial number for each element:

TABLE 3

0	I	II	III	IV	V	VI	VII	VIII		
2* He	3 Li	4 Be	5 B	6 C	7 N	8 O	9 F			
10 Ne	11 Na	12 Mg	13 Al	14 Si	15 P	16 S	17 Cl			
18 — 19 Ar	20 K	21 Ca	22 23 Sc —	24 Ti	25 V	26 Cr	27 Mn	28 Fe	29 Co	30 Ni
	31 Cu	32 Zn	33 Ga	34 Ge	35 As	36 Se	37 Br	38 —	39 —	40 —
41 — 42 Kr	43 Rb	44 Sr	45 46 Y —	47 Zr	48 Nb	49 Mo	50 —	51 Ru	52 Rh	53 Pd
	54 Ag	55 Cd	56 In	57 Sn	58 Sb	59 Te	60 J	61 —	62 —	63 —
64 — 65 Xe	66 Cs	67 Ba	68 69 La —	70 Ce	71 Nd	72 Pr	73 —	74 Sa	75 Eu	76 Gd
	77 Tb	78 (Tb$_2$)	79 Dy	80 Ho	81 Er	82 Ad	83 AcC	84 TuI	85 TuII	86 AcA
87 — 88 AcEm	89 AcX	90 TuIII	91 92 RAc Cp	93 Ct	94 Ta	95 Wo	96 —	97 Os	98 Ir	99 Pt
	100 Au	101 Hg	102 Tl	103 Pb	104 Bi	105 RaF	106 ThC	107 RaC	108 ThA	109 RaA
110 ThEm 111 RaEm	112 ThX	113 Ra	114 115 RTh Io	116 Th	117 UII	118 U	119 —	120 —	121 —	122 —

* Atomic number.

Figure 3.4 Van den Broek's periodic table of 1913.

> The serial number of every element in the sequence ordered by increasing atomic weight equals half the atomic weight and therefore the intra-atomic charge.[7]

This article was cited by no less a person than Niels Bohr in his famous trilogy paper of the same year, 1913, the work that traditionally marks the birth of the quantum theory of the atom.[8]

Nevertheless, the *most* significant development took place in a follow-up article that van den Broek placed in *Nature* magazine in which he jettisoned the connection with atomic weight altogether. After all, the Rutherford-Barkla formula only holds for elements with low atomic weights and becomes progressively less appropriate for heavier ones. Van den Broek's particular modus operandi becomes clear if one analyzes how he was able to make the necessary disconnection between atomic number and atomic weight, as will be seen in the following section.

GOODBYE TO ATOMIC WEIGHT, WELCOME ATOMIC NUMBER

The way in which van den Broek succeeded in divorcing his atomic number concept from atomic weight lies at the heart of his original contribution. I believe that he was ideally placed to do this because of his training in econometrics and his interest and ability in manipulating numerical data. In any case it is something that the physicists of the day did not choose to do, otherwise one of them might easily have made this important discovery.

It began with a set of experiments by Hans Geiger and Ernest Marsden which aimed at examining the ratio of scattering of α-particles per atom in several elements (fig. 3.5).[9] According to Rutherford this ratio needed to be constant but this is not what

I	II	III	IV	V	VI
Substance	Air equivalents of foils used	Total number of scintillations Counted for each substance	Number N of scintillations at same angle and for same air equivalent	$A^{3/2}$	$N/A^{3/2}$
Gold..........	.52, .68	1200	2400	2770	.85
Platinum...	.54, .625	1000	2900	2730	1.08
Tin.............	.51, 1.51	1400	1290	1300	.99
Silver.........	.38, .435	600	1060	1120	.95
Copper.......	.495, .61	1300	570	507	1.12
Aluminium.	.45, .52, 1.06	1600	151	144	1.05
Carbon.......	.55, .57	400	57	41.6	1.37

Figure 3.5. Geiger and Marsden. Ratio of scattering of α-particles per atom in several elements.

Geiger and Marsden found when they divided the scattering by the atomic weight of each element. Nevertheless, Geiger and Marsden were not unduly perturbed by this discrepancy, believing that the error was small enough to ignore.

Meanwhile van den Broek, the "numbers man," set to work trying to make the ratio more constant. He did this by dividing the amount of scattering for each element by its charge rather than by its atomic weight. As a result he found that the ratio was much closer to being a constant and this reinforced the notion that charge was a more important criterion for identifying any element than atomic weight.

Let me be more specific. In a section of their 1913 article, Geiger and Marsden investigate the variation of α-particle scattering with the atomic weight of a handful of elements. For example, they report the following values (fig. 3.6) for the number of scintillations per

Element	A	$N / A^{3/2}$
Al	27.1	0.24
Cu	63.57	0.23
Ag	107.88	0.18
Sn	119	0.21
Au	197.2	0.21

Figure 3.6. Rather than the representation chosen by Geiger and Marsden, the order of the elements have been inverted to show increasing atomic weights from top to bottom. Condensed by the author.

Element	scattering per atom/A^2	
	Experiment 1	Experiment 2
Cu	3.7	3.95
Ag	3.6	3.4
Sn	3.3	3.4
Pt	3.2	3.4
Au	3.4	3.1

Figure 3.7. Geiger and Marsden's experiments on the degree of scattering in various metals.

centimeter of equivalent air multiplied by atomic weight raised to the power of 3/2.[10] Geiger and Marsden then immediately state,

> This ratio $[N / A^{3/2}]$ should be constant according to the theory. The experimental values show a slight decrease with atomic weight.[11]

But again such disagreement between theory and experimental data does not seem to bother them too much. Furthermore, in a couple of footnotes in the same article, Geiger and Marsden also report experiments in which they have obtained the ratio of scattering per atom divided by the square of the atomic weight of an almost identical set of elements (fig. 3.7).

They then follow with

These results are similar, and indicate the essential correctness of the assumption that the scattering per atom is proportional to the square of the atomic weight, the deviations from constancy of the ratio are nearly within experimental error.[12]

These discrepancies which Geiger and Marsden regard as being due to experimental error are the starting point for two articles by van den Broek, the first in *Philosophical Magazine* and the second in *Nature*.

Van den Broek was already starting to think that charge was a better criterion for explaining questions regarding the periodic table and the structure of the atom. He was now in a position to demonstrate this fact by drawing on Geiger and Marsden's data.

First he divided the scattering per atom by the square of the charge of an atom instead of the square of its atomic weight. He thus found that the constancy predicted by Rutherford was more accurately recovered. In the same table he displayed values for scattering divided by $A^2/5.4$, where the purpose of the factor of 5.4 is simply to render the ratios based on M^2 comparable to those using A^2 (fig. 3.8).

In the second article, the same exercise is essentially repeated in a slightly more transparent manner that shows the two sets of data from Geiger and Marsden. Van der Broek's conclusion is exactly as it was in the first of these two articles, to the effect that the scattering data only agree with Rutherford's theory provided that charge is considered rather than atomic weight (fig. 3.9).[13]

Only now is van den Broek ready to take the crucial step of severing the connection between atomic weight and charge or atomic

	Cu	Ag	Sn	Pt	Au	Mean
Scattering/$(A^2/5.4)$	20.6	18.9	18.1	17.8	17.5	18.6
M	29	47	50	82	83	
Scattering/M^2	18.5	18.4	19.0	18.6	18.4	18.6

Figure 3.8. Van den Broek's scattering ratios for several metals.

Element	Mean	Mean x 5.4	Mean x A^2/M^2	M	A
Cu	3.825	20.6	18.4	29	63.57
Ag	3.5	18.9	18.4	47	107.88
Sn	3.35	18.1	19	50	119
Pt	3.3	17.8	18.7	82	195.2
Au	3.25	17.5	18.3	83	197.2

Figure 3.9. Table that appears in van den Broek's *Nature* article of November 27, 1913. The final column has been added by the author.

	C	Mg	Ar	Cr	Zn	Kr	Mo	Cd	Xe	W	Hg	U
M	6	12	18	24	30	36	42	48	54	78	84	96
P	4	10	16	22	26	32	38	42	48	54	58	70
kP^2	0	0	1	2	3	5	6	8	11	14	16	23
A(calc)	12	24	38	52	66	82	98	112	130	184	200	238
A(exp)	12	24	40	52	65	82	96	112	130	184	200	238

Figure. 3.10. Comparison of $A_{calculated}$ by van den Broek with experimental values of A as seen in final two rows. Discrepancies occur only for Ar, Zn and Mo among this set of 12 elements with a large range of atomic weights.

number. And so, atomic weight is abandoned and charge becomes the focus of attention. In his own words,

> If now in these values the number M of the place each element occupies in Mendeléeff's series is taken instead of A, the atomic weight, we get a real constant (18.7 +/- 0.3); hence the hypothesis proposed holds good for Mendeléeff's series, but the nuclear charge is not equal to half the atomic weight.[14]

PREEMPTING MOSELEY

In an article published December 25, 1913, van den Broek appears to preempt Moseley in using the concept of atomic number to predict the existence of new or missing elements. He begins

I am grateful to Mr. Soddy (*Nature*, December 4, p. 399) that in accepting in principle the hypothesis that the intra-atomic charge of an element is determined by its place in the periodic table, he directed attention to the possible uncertainty of the absolute values of intra-atomic charge and of the number of intra-atomic electrons. Surely the absolute values depend on the number of rare earth elements; but if to the twelve elements of the rare earth series, the international table contains between cerium and tantalum the new elements (at least four) discovered by Auer von Welsbach in thulium (*Monatshefte für Chemie,* 32, Mai, S. 373), further keltium discovered by Urbain (*Comptes Rendus d. l'Academie des Sciences,* 152, 141-3), and an unknown one for the open place between praeseodymium and samarium be added, this long period too becomes regular. Moreover if only twelve instead of eighteen elements existed here, the ratio of the large-angle scattering per atom divided by M^2 is no longer constant, the values of copper, silver, tin platinum and gold then being 1.16, 1.15, 1.19, 1.26, and 1.24 respectively instead of 1.16, 1.15, 1.19, 1.17, and 1.15 and the same holds for the following relation concerning the number of intra-atomic electrons.[15]

Van den Broek then proceeds to do away with the irregularities in Mendeleev's system by removing hydrogen and helium from the body of the table and placing all the group 8 elements, such as the platinum metals, into one space for each group of three elements (such as platinum, osmium, and iridium). He also derives an empirical relationship, which enables him to see even greater regularity in the lengths of his new periods as well as the ordinal number for each element and their atomic weights, thus further showing the value of concentrating upon the ordinal number. The relationship is $A_{calculated} = 2(M + kP^2)$, as M is the ordinal number in Mendeleev's table, k is an empirically obtained constant of .00468, and P is the same for this new condensed periodic system as M is for Mendeleev's system. Van den Broek proceeds to compare his calculated values of atomic weight with the experimental values, finding only a few minor discrepancies as shown in figure 3.10.

There is little doubt that Moseley would have seen articles such as this one and that they would have helped him to form his ideas about the importance of atomic number. In any case, let us not forget that Moseley begins one of his two classic articles by stating that he is conducting his experiments with the "express purpose of testing van den Broek's hypothesis."

INTERLUDE ON WHAT BOHR KNEW ABOUT ATOMIC NUMBER

Since I am making a case for the importance of van den Broek's contribution to the physics of the early 20th century it is worth spending a little time in examining just how much Niels Bohr knew or anticipated about the nature of atomic number. I am also doing this because the version of the history of physics as recounted by physicists would seem to have contributed to the downgrading of van den Broek's contribution. This is not altogether surprising given that, as I have suggested, van den Broek's approach came from an essentially chemical perspective of examining the relationship of all the elements to each other through a study of the periodic table.

In particular, John Heilbron, a preeminent historian of physics of the early 20th century has claimed, on more than one occasion, that Bohr understood atomic number regardless of the work of van den Broek.[16] I believe that Heilbron may be overstating his case, as I will try to argue. As I see it, although Bohr may have understood the importance of atomic charge as a means of identifying the number of electrons in hydrogen, and helium especially, he did not appreciate the need to separate clearly the concepts of atomic weight and atomic number.

I proceed historically by consulting the *Collected Papers of Niels Bohr* in order to trace his views on atomic number as well as his views on the experiments and a law announced by Richard Whiddington.[17] The first time that Bohr mentions Whiddington in print is in an article in the January 1913 *Philosophical Magazine* entitled "On the Theory of the Decrease of Velocity of Moving Electrified Particles." Here Bohr discusses Thomson's theory on the passage of α and β particles through matter and how their velocity decreases in the process. He also mentions that Whiddington's experiments have confirmed Thomson's theoretical work on this subject.[18] Later in the same article Bohr mentions an earlier work by Whiddington whereby the minimum velocity of a particle required to excite the characteristic X-radiation from any target element is proportional to the atomic weight of the element.[19] More precisely,

$$v_{minimum} = 10^8 \times A \, m/sec$$

Bohr proceeds to apply Whiddington's law in the following way. He proposes that the energy of such a particle (kinetic energy $= 1/2mv^2$) should therefore be,

$$Energy = (m/2) \times 10^{16} \times A^2$$

He then invokes Planck's radiation law which he gives in the form of

$$E = v \times k$$

where v is the number of vibrations per second in a Planckian atomic vibrator and k is a constant equal to 6.55×10^{-27}. Equating these two expressions for energy Bohr obtains

$$v \times k = (m/2) \times 10^{16} \times A^2$$

and solving for v after substituting for the values of k and m gives

$$v = 6.7 \times 10^{14}$$

He then considers the case of an oxygen atom, substituting the atomic weight for this element or $A = 16$ to obtain

$$v = 1.7 \times 10^{17}$$

and finally for the frequency of vibration n, using the simple expression

$$n = 2\pi v$$

he obtains

$$n = 1.1 \times 10^{18}$$

As Bohr states, this value bears "remarkable" agreement with the value for the vibrational frequency calculated by a quite separate approach from the absorption of a-rays, namely 0.6×10^{18}.

But for elements heavier than oxygen the agreement is rather less remarkable. Using the measured velocity of particles passing through a number of other elements, Bohr calculates a quantity that he calls r which he states is about half of the atomic weight of each element, according to Rutherford's theory. This is of course the approximate relationship $Z \approx A/2$ which has already appeared numerous times in this chapter, but using r in place of Z. Bohr finds the following results (fig. 3.11):

Element	Atomic weight	A/2	r
Aluminium	27	13.5	14
Tin	119	59.5	38
Gold	197	98.5	61
Lead	207	103.5	65

Figure 3.11. A table compiled by present author based on a table given on page 27 of Bohr's article.

Bohr writes,[20]

According to Rutherford's theory we shall expect values for r equal to half of the atomic weight; we see that this is the case for aluminium, but that the values of r for the elements of higher atomic weight are considerably lower.

He then mentions corrections that would need to be made, all the time seemingly believing that r must eventually be made to equal $A/2$.[21] My purpose for belaboring this point is to show that at this stage Bohr had no inkling that Z and A would have to be sharply distinguished from each other. Like Geiger and Marsden, as mentioned earlier, Bohr does not seem to show any great concern for this discrepancy between the experimental finding and the relationship $Z \approx A/2$ as it applies to elements with high atomic weights. On the other hand this discrepancy was the crucial feature that motivated van den Broek to eventually unravel the question satisfactorily by severing the connection between Z and A, since the relationship only applies to elements with low atomic weights.[22]

Nevertheless, in the concluding section of this article Bohr can claim with some confidence that hydrogen has one electron while helium has two. About elements with higher atomic weights, all he can say is,[23]

For elements of higher atomic weight, it is shown that the number and frequencies of the electrons, which we must assume, according to the theory, in order to explain the absorption of α-rays are of the order of magnitude to be expected.

As in the case of Geiger and Marsden, agreement within an order of magnitude seems to be good enough for Bohr. Contrary to Heilbron, I do not believe that this work suggests that Bohr had a good understanding of the notion of atomic number. That honor I insist belongs to van den Broek alone.

CONCLUSION

The second intermediate character in this book differs from many others that will be discussed here in being a complete outsider to the scientific establishment of the day. This is one of the reasons for van den Broek's neglect, but it cannot fully explain the fact that he remains virtually unknown even within histories of atomic physics. The evolution in his thinking, which we have examined here, serves to highlight rather well the organic manner in which scientific developments seem to propagate. While en route to the discovery of atomic number, van den Broek entertained many ideas which in retrospect were plain wrong. But such ideas as the existence of the alphon particle and the need for a three-dimensional periodic table appear to have aided his thought process. Just like Nicholson, whose apparently shaky foundations led to progress, so van den Broek was able to outpace his professional colleagues like Bohr, Rutherford, Barkla, and many others. Only van den Broek clearly articulated the need to jettison atomic weight as an ordering principle for the elements and to replace it with the concept of atomic number that he identified with the charge on the nucleus of any particular atom. The "logic of discovery," as some wish to call it, seems to follow a path that is far from logical. I nominate van den Broek as one of my seven missing links in the trajectory of how atomic theory developed in the 20th century.

NOTES

1. E.R. Scerri, *The Periodic Table, Its Story and Its Significance*, Oxford University Press, New York, 2007.

2. E.R. Scerri, "The Discovery of the Periodic Table as a Case of Simultaneous Discovery," *Philosophical Transactions of the Royal Society A*, 373, 20140172, 2015.

3. A. van den Broek, "Das α-Teilchen und das periodische System der Elemente," *Annalen der Physik*, 4, 23, 199–203, 1907, quotation is from p. 199.

4. A. van den Broek, *Physikalische Zeitschrift*, xii, 490–497, 1911, quotation is from p. 491.

5. A. van den Broek, "The Number of Possible Elements and Mendeleeff's 'Cubic' Periodic System," *Nature*, 87, 78–78, 1911.

6. A. Pais, *Inward Bound*, Oxford, Clarendon Press, 1986, p. 227.

7. A. van den Broek, "Die Radioelemente, das periodische System und die Konstitution der Atome," *Physikalische Zeitschrift*, 14, 32–41, 1913.

8. N. Bohr, "On the Constitution of Atoms and Molecules" [in three parts], *Philosophical Magazine*, series 6, 26, 1–25 [I], 476–502 [II], and 854–875 [III].

9. H. Geiger and E. Marsden, "The Laws of Deflection of α-Particles through Large Angles," *Philosophical Magazine*, xxv, 604–628, 1913, quotation is from p. 604.

10. The factor of $3/2$ arises as follows: The scattering of α-particles is proportional to the square of the atomic weight of any element. Moreover, for different foils of the same air equivalent the number of atoms per unit area is inversely proportional to the square root of atomic weight. Combining these two relations, $A^2 \times A^{-1/2} = A^{3/2}$.

11. H. Geiger, E. Marsden, "The Laws of Deflection of α-Particles through Large Angles," *Philosophical Magazine*, xxv, 604–628, 1913. Based on table IV on p. 617.

12. Ibid., p. 619.

13. A. van den Broek, "Inter-Atomic Charge and the Structure of the Atom," *Nature*, 92, 372–373, 1913.

14. Ibid. Table appears on p. 373.

15. A. van den Broek, "Inter-Atomic Charge and the Structure of the Atom," *Nature*, 92, 476–478, 1913, quotation is from p. 476.

16. Public lectures by J. Heilbron, in Manchester, UK, October, 2013. https://www.youtube.com/watch?v=EYIhx5YFNgo

17. R. Whiddington, "The Production of Characteristic Röntgen Radiations," *Proceedings of the Royal Society*, 85, 323–332, 1911.

18. Ibid.
19. Ibid.
20. N. Bohr, "On the Theory of the Decrease of Velocity of Moving Electrified Particles on Moving through Matter," *Philosophical Magazine*, 25, 10–31, 1913, quotation is from p. 27.
21. To say, as Bohr does, that corrections are needed is to state the obvious. The remedy is rather to disconnect the charge of the nucleus from atomic weight as van den Broek carried out.
22. The approximate relationship applies only to the first 20 or so elements, from hydrogen to calcium, after which it shows an increasing divergence as Z and A increase.
23. N. Bohr, "On the Theory of the Decrease of Velocity of Moving Electrified Particles on Moving through Matter," *Philosophical Magazine*, 25, 10–31, 1913, quotation is from pp. 30–31.

Chapter 4

Richard Abegg

A BRIEF BIOGRAPHY AND INTRODUCTION

Richard Abegg (fig. 4.1) is one of several chemists featured in this book. He was born in 1869 in Danzig, which is today part of Poland, in the same year that Mendeleev published his first periodic table. Although his family was of Swiss origin, Abegg was raised in Germany, attending elementary and secondary school in Berlin; even as a boy, he kept a chemical laboratory in his home, much to the annoyance of his parents. As a university student he had a succession of illustrious teachers including Lothar Meyer in Tubingen, Ladenburg in Kiel, and A. W. Hoffman in Berlin. After graduating with a bachelor's degree Abegg continued his studies with Ostwald in Leipzig, then moved on to Svante Arrhenius in Stockholm in 1892–93.

In 1894 Abegg became Privatdozent and an assistant to Walter Nernst in Gottingen, moving to a position in Breslau in 1901. Although initially trained as an organic chemist Abegg switched to physical chemistry and became an avid supporter of the views of Arrhenius, van't Hoff, and Ostwald. The influence of these mentors is evident in several books that Abegg wrote in which he frequently expressed the view that chemistry should be elevated from a descriptive science into a rational one.

By 1900 Abegg was serving in the German army where he made the first of several balloon flights. He appears to have been so taken by these experiences that he continued to take balloon trips for pleasure, only to die prematurely in a ballooning accident at the age of just 41 while at the height of his powers.

Figure 4.1. Richard Abegg.

The work by Abegg that most concerns us here was begun in 1899 with an article under the title of "Valency and the Periodic System." It was an electrochemical interpretation of a relationship between the elements that had been published by Mendeleev about 30 years previously and which has become known as Mendeleev's rule of eight. Abegg was to give a deeper interpretation of this rule than Mendeleev's, and this served as a bridge to the earliest attempts to explain chemical periodicity in electronic terms.

As in the case of the other personalities in this book, the pivotal role is widely accepted in the most accurate historical accounts, but almost universally denied in the more popular imagination as well as in textbook accounts—even those intended for specialists. If we ignore such intermediate steps, the development of science appears to happen as a series of spectacular leaps and the architects of successive theories take on the mantle of superheroes with almost magical powers of divination. The more correct picture, I suggest, is one of incremental steps occurring almost imperceptibly and frequently carried out by unknown individuals. The history of science proceeds via evolution in which dozens of small players contribute and not via revolutions fashioned by the few and famous.

ABEGG'S EARLY WORK IN ELECTROCHEMISTRY

Abegg's work on the theory of valency occurred as a direct consequence of his study and apprenticeship with the early pioneers of theories of electrochemistry and electrical theories of chemical bonding. The second of these branches in particular has a long and varied history that goes back to the work of the chemists Davy and Berzelius, and even further back to Dalton's revival of the atomic theory.

John Dalton believed that there was a natural attraction between unlike atoms and that there was repulsion between atoms of the same element. The Swedish chemist Jacob Berzelius began by supporting Dalton's atomic theory although he soon departed from him in several important ways. For example, Berzelius took seriously the law of combining gases, as discovered by Gay-Lussac and Humbolt, which held that gas volumes react together to give gaseous products whose volumes stood in a simple ratio to each other. This led Berzelius to realize that water had a formula of H_2O rather than HO as Dalton had assumed.[1]

Berzelius also went beyond Dalton in postulating a theory to account for the affinity of unlike atoms. He believed that such affinity was of an essentially electrical nature. Berzelius conducted many of the early electrolysis experiments that became possible after Volta invented the electrical battery in the very early 1800s. Berzelius reasoned that since chemical combination in a battery produced electricity and since this in turn caused chemical decomposition, such as electrolysis, one should conclude that chemical compounds were held together by electrical forces.

> Dans toute combinaison chimique il y a neutralisation des électricités opposées, et cette neutralisation produit le feu, de la même manière qu'elle le produit dans les décharges de la bouteille électrique, de la pile électrique, et du tonnerre… Si les corps, qui se sont unis, on cessé d'être électrique, doivent être sépare… il faut qu'ils recouvrent l'état électrique détruit de la combinaison. Aussi sait-on par l'action de la pile galvanique sur un liquide conducteur, les éléments de ce liquide se séparent, que l'oxygène et les acides se rendent du pole négatif au pole positif, tandis que les corps combustibles et les bases salifiable sont poussés du pole positif au négatif.[2]

Berzelius therefore proposed that chemical affinity is a manifestation of electrical attraction and consequently, that atoms of the elements must have an electrical character.

> The atoms of each elementary body have 2 poles, on which two opposite electricities are accumulated in different proportions, according to the nature of the bodies: for example, oxygen has a large negative quantity, and a small positive quantity, while potassium has the opposite disposition.[3]

Berzelius also compiled one of the earliest lists of elements in the form of an electrochemical series in which he regarded oxygen as being the most negative and potassium as the most positive. The list bears a striking similarity to the modern electrochemical series that continues to serve a very useful purpose in modern chemistry.[4]

His dualistic theory proposed that all compounds should be divisible into two parts with opposite electrical charges. The idea would prove very effective in inorganic chemistry but would meet its greatest challenge in the field of organic chemistry. Berzelius's theory fell into disrepute by the 1830s when it was found that hydrogen, which according to Berzelius was an electropositive element, could be substituted by the electronegative element of chlorine in several organic compounds. While this should not be possible according to Berzelius's dualistic theory, experiments were showing that it did indeed happen.

After a considerable hiatus in the study of electrical approaches to bonding, such theories sprang up again in various guises, with many seemingly isolated tributaries building up to a main stream by the end of the 19th century. Among these theories, and one that was to have a special influence on Abegg, was the theory of ionic dissociation as proposed by Arrhenius.[5] In simple terms the theory proposed that during electrolysis the current is carried by tiny electrically charged ions that move at different speeds. The electric current could

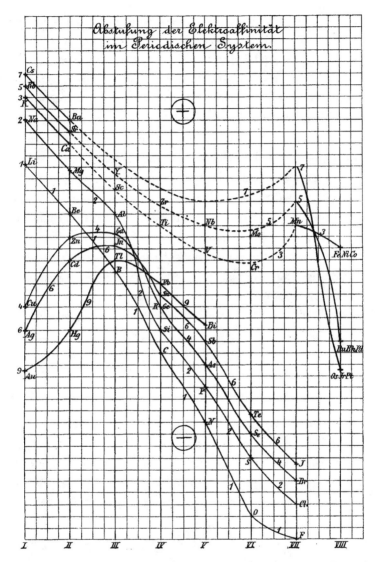

Figure 4.2. A diagram from Abegg's article of 1904 showing an element's electroaffinity as a function of its position in the 8-column periodic table.

be regarded as propelling the oppositely charged ions such as Na^+ and Cl^- in opposite directions toward their oppositely charged electrodes.

In 1879 Abegg and Bödlander began to publish articles on their theory of electroaffinity. The general idea was to use a quantity called the half-cell potentials of any element (as explained in the next section) as a measure of their attraction for electrons. Moreover, they attempted to use the electroaffinity of the elements in order to examine periodic trends, such as solubility and the ability of elements to form complex ions.[6]

Abegg coauthored a seminal contribution with Guido Bodlander in 1899 that acted as a bridge between his earlier interests in electrochemistry and ionic theories and the study of the periodic table. In this article the authors proposed a theory of what they termed electroaffinity, whereby measured values of half-cell oxidation potentials, for any particular metal/aqueous metal ion system, were taken to be a measure of an atom's attraction for its electrons.[7]

We should pause to explain some of these terms, for readers who may not be sufficiently well versed in chemistry.

WHAT IS A HALF-CELL?

A half-cell consists of a metal dipping into a solution of its own aqueous ions. For example one may think of a piece of pure zinc that has been placed inside a molar solution of aqueous ions of Zn^{2+}. In such cases two competing processes occur on a microscopic level. The first of these is a tendency for the metal cations to leave the metal and to enter into the aqueous solution. This process leads to a buildup of negative charge on the solid metal. At the same time, there is an opposite tendency whereby aqueous ions of Zn^{2+} deposit themselves onto the metal surface and in so doing cause an increasingly positive charge to develop on the piece of metal. For any particular metal–metal ion system of this kind, the extent of the two tendencies differ, with the result that the piece of metal will adopt either a negative or positive net charge. In the case of most metallic systems of this kind, a negative charge is adopted.

The charge or potential of any such system can be measured relative to a hydrogen electrode which is arbitrarily assigned a value of exactly zero volts. These values are half-cell or also called oxidation potentials, since they provide a measure of the oxidation of any metal relative to its metal ions which possess a higher oxidation state than the uncombined metal.[8]

So what Abegg and his coauthor were assuming was that the oxidation potential of any metal was directly correlated with the ability of a metal to lose electrons. With the benefit of hindsight it becomes clear that such a simplistic correlation is incorrect. The energy changes involved in converting a solid metal atom into an aqueous ion is now known to involve two additional kinds of changes, namely the energy of atomization of a metal to form a gaseous atom and hydration energy to convert a gaseous atom ion into an aqueous ion. Rather than equating oxidation potentials with the ability of an atom to lose electrons, one should rather equate oxidation potential with the sum of three energetic changes,

Oxidation potential = atomization + ionization + hydration energy

The relevance of this more accurate treatment is nowhere better seen than on considering the sharp change which takes place in oxidation potentials as one moves from the metal copper to the subsequent metal in the periodic table, namely zinc. This also highlights why simplistic correlations between oxidation potentials and periodic trends can easily break down. For example, whereas the value of oxidation potential increases fairly regularly as one moves through the first transition series of elements, there is a sudden drop in value from that of copper to that of zinc, two successive elements in the periodic table. The cause of this abrupt change lies in zinc having an anomalous value for its atomization energy while the ionization energies of the two metals are rather similar.[9]

While completely unaware of these subtleties, Abegg and Bodlander forged ahead and tried to correlate a number of periodic properties with values of measured oxidation potentials. They were immediately criticized, for quite separate reasons than those mentioned above, by one John Locke, a chemist from Yale University. Among many specific points Locke wrote,[10]

The solubilities of the salts of magnesium, manganese, ferrous iron, nickel, cobalt, copper, and cadmium are qualitatively almost the same as those of zinc compounds, though in the matter of decomposition voltage (oxidation potential) the metals range from +1.47 to −0.34. Here in the matter of solubility we have quantitative data, which can be compared directly with the electroaffinities, or with the decomposition voltage values, which roughly measure the latter [fig. 4.2]. The solubilities of the nitrates, $M(NO_3)_2.6H_2O$, at 18°, expressed in gram-molecules per liter of solvent are as follows,

	Mg	Mn	Fe	Co	Ni	Cu	Zn	Cd
Solubilities	4.95	7.41	4.57	5.40	5.12	6.13	6.08	5.37
Voltage	+1.47	+1.06	+0.33	+0.22	+0.22	−0.34	+0.74	+0.38

Figure 4.3. Excerpt from John Locke's critique of Abegg and Bodlander.

But as in many cases encountered in this book, there was enough that was incrementally advantageous in Abegg's approach that it reinforced his belief in a strong connection between electrical phenomena such as electroaffinity, that he had introduced, and chemical properties of the elements and the periodic system as a whole (fig. 4.4).

ABEGG ON VALENCY AND THE PERIODIC TABLE

Mendeleev's "rule of eight" was the precursor of much that was to follow in the study of electronic structure. One might almost say that Mendeleev anticipated Lewis's octet rule or even that he anticipated the notion of a full octet of electrons.[11] But we are not in the business of establishing anticipations but rather arguing for evolution and continuity in the history of science, meaning that each of these discoveries led into another one through an organic evolutionary process. Rather than being puzzled by possible anticipations we will try to view the development of such ideas from within, as it were, and see this as a perfectly natural process.

Mendeleev produced different versions of his rule of eight. One version appears in his textbook, while he was in the process of revising the laws of valency that he felt were in need of reform. In this section he lists three principles of which the third is the one of interest for our purposes. Under the subtitle of "The Periodic Principle," Mendeleev writes,

> The highest compounds of an element with hydrogen, oxygen and other equivalent elements are determined by the atomic weight of the element, of which they form a periodic function.

Hydrides provide only four forms, namely RH, EH_2, RH_3, and EH_4. Meanwhile, with oxygen the following forms are found,

$$R_2O, \quad RO, \quad R_2O_3, \quad RO_2, \quad R_2O_5, \quad RO_3, \quad R_2O_7, \quad RO_4$$

Mendeleev notes that for no element does the sum of the hydrogen and oxygen equivalences exceed eight.

Series.	Group I. R₂O.	Group II. RO.	Group III. R₂O₃.	Group IV. RH₄. RO₂.	Group V. RH₃. R₂O₅.	Group VI. RH₂. RO₃.	Group VII. RH. R₂O₇.	Group VIII. RO₄.
1 ········	H=1							
2 ········	Li=7	Be=9.4	B=11	C=12	N=14	O=16	F=19	
3 ········	Na=23	Mg=24	Al=27.3	Si=28	P=31	S=32	Cl=35.5	
4 ········	K=39	Ca=40	—=44	Ti=48	V=51	Cr=52	Mn=55	Fe=56, Ce=59 Ni=59, Cu=63
5 ········	(Cu=63)	Zn=65	—=68	—=72	As=75	Se=78	Br=80	
6 ········	Rb=85	Sr=87	? Y=88	Zr=90	Nb=94	Mo=96	—=100	Ru=194, Rh=104 Pd=106, Ag=108
7 ········	(Ag=108)	Cd=112	In=113	Sn=118	Sb=122	Te=125	I=127	
8 ········	Cs=133	Ba=137	? Di=138	? Ce=140	....	....	....	....
9 ········	....	....	....	....	....	....	....	....
10 ········	....	....	? Er=178	? La=180	Ta=182	W=184	....	Os=195, In=197 Pt=198, Au=199
11 ········	(Au=199)	Hg=200	Tl=204	Pb=207	Bi=208	....	....	....
12 ········	....	....	....	Th=231	....	U=240	....	....

Figure 4.4. One of the most popular versions of Mendeleev's periodic table, as published in 1871.

Oxide	RO_4	R_2O_7	RO_3	R_2O_5	RO_2
Hydride	none	RH	RH_2	RH_3	EH_4

Significantly, for what is to come, this relationship seems to only apply to elements from just four groups in the periodic table. Moreover, Mendeleev seems to consider this relationship sufficiently important to the point of embedding it into the headings for each of the groups as shown in his table in figure 4.4.

Another version of Mendeleev's rule of eight was announced in his most substantial article on the periodic table, in which he made the famous predictions of new elements as well as correcting the placement of certain elements and correcting the atomic weights of yet other elements.[12]

Here is the relevant passage from Mendeleev's article:

So far our manner of looking at things differs from that adopted by the partisans of valency theory only in its outer appearance, but is identical in its essentials because it is based only on hydrogen and its compounds. Later we shall meet with essential differences. Likewise, it should be noted that only a few elements are capable of giving hydrogen compounds, especially as homologs and unsaturated substances. So far as we know, only carbon gives them in any great number.

Oxygen combines with each element to produce one of the following forms:

$$R_2O, RO, R_2O_3, RO_2, R_2O_5, RO_3, R_2O_7, RO_4$$

.... Judging from the constitution of water, H_2 and O are equivalent to the hydrogen forms,

$$RH, RH_2, RH_3, RH_4$$

Because the sum of the number of equivalents of hydrogen and oxygen which can be bound to one atom of the element does not exceed eight, elements which give RO_4 do not form compounds with hydrogen.[13] Those elements which can give R_2O_7 give RH; those which give RO_3, give RH_2; those which give R_2O_5, give RH_3; and those which give RO_2, give RH_4. Elements corresponding to the highest form R_2O_3, have not as yet given any compounds with hydrogen, because no additional hydrogen form RH_5 exists.[14]

Due to these apparent limitations, noted in the final sentence of this passage, Mendeleev did not include any hydrogen compounds in the first three columns of his periodic table of 1871 when giving the formulas at the head of each column (fig. 4.4).

Figure 4.5 provides a clearer way to see what Mendeleev is proposing.[15] The first row lists the oxides of elements in the sodium period. The second row lists the hydrides of the same elements. Whereas the respective oxygen and hydrogen compounds from silicon onwards obey his rule of eight, those of the first three elements do not, given that the sum of the combining atoms of oxygen and hydrogen are 2, 4, and 6 respectively rather than 8.

Groups	I	II	II	IV	V	VI	VII
	Na_2O	Mg_2O_2	Al_2O_3	SiO_2	P_2H_5	S_2O_6	Cl_2O_7
	NaH	Mg_2O_2	AlH_3	SiH_4	PH_3	SH_2	ClH
	2	4	6	8	8	8	8

Figure 4.5. Table constructed using Mendeleev's examples, with valences added by the author in the lowest row.

Group	1	2	3	4	5	6	7
Normal valence	+1	+2	+3	+/–4	–3	–2	–1
Contra valence	–7	–6	–5	+/–4	+5	+6	+7

Figure 4. 6. Abegg's normal and contravalences that together add to eight (ignoring the signs in front of each individual value).

It can be seen that Mendeleev's rule of eight is of somewhat limited applicability. This point is important for what is to follow concerning Richard Abegg's contribution, since he was able to make the rule more general by means of a simple maneuver. Instead of confining himself to compounds of oxygen and hydrogen, Abegg concentrated on the maximum and minimum valences available to each element, even if this was the case only in principle (fig. 4.6). As Jensen has explained, the deeper significance of Abegg's approach is that he does not face the same problem as Mendeleev when it comes to compounds from groups I to III in the periodic table.[16] It is as though Mendeleev was assuming that hydrogen was the most electronegative element whereas in fact elements such as those in groups I to III are typically more electropositive than hydrogen. Of course Mendeleev could not approach matters from an electrical point of view, as Abegg did, since no such electrical views had yet been developed in the 1870s. And even when such views were developed, by the likes of Arrhenius, Mendeleev remained famously opposed to them.

On the other hand, as we have seen, Abegg had obtained an expert knowledge of electrochemistry that allowed him to correct Mendeleev's more restricted view.

According to Abegg's rule of eight, elements are in principle capable of showing a maximum electropositive valence (normal valence) and a maximum electronegative valence (contravalence) in which the sum of the two valences is always equal to 8. Clearly Abegg had succeeded in generalizing Mendeleev's rule, in making it more abstract and in removing the apparent problems that had prevented Mendeleev's rule from being applicable to all eight groups of the periodic table. It would be a short step from Abegg's rule of eight to Lewis's octet rule to explain chemical bonding in terms of elements striving to obtain a full outer shell of electrons.

Comparison between Mendeleev and Abegg's rules:

$$\text{Mendeleev } v(O) + v(H) \leq 8 \text{ for groups IV to VIII}$$

$$\text{Abegg } v(+) + v(-) = 8 \text{ for groups I to VIII}$$

CONCLUSION

Once again, Abegg provides an example of a little-known scientist establishing some missing links. In this case it was between the work of Mendeleev on valency and G. N. Lewis's pioneering ideas on chemical bonding in terms of numbers of electrons. Finally, this claim is highlighted by considering the following quotation from Abegg's article of 1899,

> The sum of eight of our normal and contra-valences has there-fore the simple significance as the number which represents for all the atoms the points of attack of electrons; and the group number of positive valency indicates how many of the eight points of attack must hold electrons in order to make the element electrically neutral.[17]

This statement is one of the earliest expressions of the role that electrons play in chemical bonding. Abegg's points of attack would soon become the eight electrons arranged at the corners of cubes, which would it turn become Lewis's pairs of electrons at the corners of a tetrahedron and eventually a ring of eight electrons.

Abegg belongs firmly in the historical development that flowed in an organic fashion from Mendeleev through to G. N. Lewis and eventually other pioneers of theories of chemical bonding like Linus Pauling. Science historians generally regard this as a sequence of individual contributions and of course this is true. But over and above "who did what when," there is the undeniable fact that the body-scientific provides an ever more sophisticated and smoothly linked growth in ways of thinking of chemical combination between different elements.

NOTES

1. E.R. Scerri, *The Periodic Table, Its Story and Its Significance*, Oxford University Press, New York, 2007.

2. *La théorie des proportions chimiques et de l'influence de l'électricité dans la nature inorganique*, Authorized French Edition, Méquignon-Marvis, Paris, 1835, p. 46. Author's translation: A neutralization of opposite charges occurs in all combinations and this neutralization produces a spark in the same way as is produced in a discharge jar, an electric pile, or in lightning. If the bodies that are united cease being charged and need to be separated, they must recover the electrical state that is removed by their combination. We also know from the action of a galvanic pile, on a conducting liquid, that the elements of this liquid separate, that oxygen and acids are deposited on the negative pole, whereas combustible bodies and basic salts are repelled from the positive to the negative pole.

3. J. Berzelius in Gmelin's *Handbuch der Chemie*, translation by Watts, Cavendish Society, London, 1848.

4. "Electrochemical Series," In W.M. Haynes, *Handbook of Chemistry and Physics: 93rd Edition*, Chemical Rubber Company, pp. 5–80.

5. This theory in turn was a development of the earlier work of Faraday and Helmholz, both of whom favored the use of electrical notions in the study of chemistry.

6. Jensen suggests that Abegg's use of electroaffinity might be analogous to the way in which contemporary chemists use the concept of electronegativity in order to rationalize numerous periodic trends.

7. The equivalent modern notion is termed electronegativity, of which there are many versions and the best known of which is due to Linus Pauling. L. Pauling,

"The Nature of the Chemical Bond. IV. The Energy of Single Bonds and the Relative Electronegativity of Atoms," *Journal of the American Chemical Society* 54 (9), 3570–3582, 1932.

8. Modern work in this area is almost invariably discussed in terms of reduction potentials which measure the opposite process whereby a metal ion is reduced to its neutral metallic state. I choose to work with oxidation potentials in this account since this is what Abegg was using throughout his studies.

9. S.H. Strauss, "In Search of a Deep Understanding of $Cu^{2+/0}$ and $Zn^{2+/0}$ E^0 Values," *Journal of Chemical Education*, 76, 1095–1098, 1999.

10. J. Locke, "Electro-Affinity as a Basis for the Systematization of Inorganic Compounds," *American Chemical Journal*, 27, 105–117, 1902.

11. Mendeleev did not call it a rule and nor did he draw a great deal of attention to it. *Mendeleev on the Periodic Law, 1896–1905*, selected and edited by W.B. Jensen, Dover, Mineola, NY, 2005.

12. Ibid.

13. As a matter of fact of these elements only osmium and iridium do not form hydrides.

14. Mendeleev quoted from p. 102 of *Mendeleev on the Periodic Law, 1896–1905*, selected and edited by W.B. Jensen, Dover, Mineola, NY, 2002.

15. This table does not appear in any of Mendeleev's articles or those of Abegg. It has been constructed by the present author with a view to showing the limitations of Mendeleev's rule as directly as possible. The usual formulas for some compounds such as H_2S and HCl have been inverted for the sake of comparison.

16. W.B. Jensen, "Abegg, Lewis, Langmuir and the Octet Rule," *Journal of Chemical Education*, 61, 191–200, 1984.

17. R. Abegg, "Die Valenz und das periodische System: Versuch einer Theorie der Molekülarverbindung," *Zeits Anorganische Chemie*, 39, 330, 1904, quotation is from p. 380.

Chapter 5

Charles Bury

Charles Bury is the second transitional figure to be discussed in this book who was primarily a chemist (fig. 5.1). Nevertheless he contributed a key piece of work that lies at the border of chemistry and physics, or more specifically the elucidation of the electronic configurations of the atoms of the elements. Nor was he the first chemist to participate in this endeavor, being preceded by G. N. Lewis and Irving Langmuir, both of whom are far better known.

Although Lewis was never awarded a Nobel Prize he is one of the best-known chemists of the 20th century due to his enormous contributions to the theory of chemical bonding. For example, it was Lewis who first arrived at the now universal notion that a chemical bond consists of a shared pair of electrons. This notion has been somewhat superseded in the later quantum mechanical theories of bonding and yet remains as a very useful visualizeable notion in both chemical education and professional chemistry.

Langmuir was awarded the 1932 Nobel Prize in chemistry for developments in surface chemistry including that of oil films. In addition, Langmuir did a great deal to popularize Lewis's theories of chemical bonding: For example, although the concept of the octet rule, whereby atoms bond in such a way as to achieve a full octet of electrons was originally due to Lewis, it was Langmuir who made it widely known.[1]

To better understand the work of Charles Bury, by far the least well known of these chemists, it will be necessary to begin with a review of the electronic configurations of atoms as understood by physicists

Figure 5.1. Charles Bury.

Thomson, Bohr, and Kossell as well as the chemists Lewis and Langmuir. Before conducting this survey a few more general comments about the assignment of electronic configurations according to chemists as well as physicists are also in order.

It is no exaggeration to say that in general the configurations arrived at by Lewis, Langmuir, and Bury were superior to those of physicists such as Bohr, Stoner, and Pauli, all of whose work will be treated in later chapters. The reason for this difference in the configurations is not difficult to appreciate. Whereas physicists were working deductively from general principles, they were forced to adopt a semiempirical approach of consulting experimental data such as spectral and chemical evidence. Conversely, the chemists made no pretense of deriving configurations from general principles but helped themselves to empirical information regarding how each of the various atoms bonded to other atoms. And in this last respect they were naturally ahead of the physicists, in that they had a better grasp of how each element forms chemical bonds.

The notion of an electronic arrangement or configuration, to use the current term, was certainly not introduced by Bury. It has its origins in the independent work of a number of physicists and chemists and provides yet another example of the gradual evolution of a scientific concept—and therefore of one of the main theses in this book.

Electronic configurations developed along more or less independent lines among chemists and physicists, although they eventually merged, in order to explain the main features, as well as the details, of the periodic table. But even before such a merging took place the chemists and physicists involved were motivated by a desire to explain the existence of the periodic law and its graphical representation in the form of the periodic table.

ELECTRONIC CONFIGURATIONS ACCORDING TO PHYSICISTS

Thomson

On the side of the physicists, we begin with J. J. Thomson, who had discovered the electron in 1897.[2] Thomson then set about trying to explain the periodic table, a feat that he declared to be one of the major goals of theoretical physics at the time. Thomson's approach was to consider the mechanical stability of rings of electrons circulating within the positive charge of his "plum pudding" atom, rather than attending to chemical behavior as Lewis had done.[3] Perhaps this feature accounts for why Thomson's configurations were less successful in trying to explain chemical periodicity.

Thomson concluded that the early solar system-like models of atoms proposed by Perrin and Nagaoka would be unstable because the orbiting electrons would continuously radiate energy, eventually falling into the center of the atom. Instead he suggested an alternative model in which the electrons were embedded in the nucleus, circulating within its positive charge or became known as his "plum pudding" model of the atom. In the same paper of 1904 Thomson also

Number of electrons	Rings	Number of electrons	Rings
5	5	16	5 + 11
6	1 + 5	17	1 + 5 + 11
7	1 + 6	18	1 + 6 + 11
8	1 + 7	19	1 + 7 + 11
9	1 + 8	20	1 + 7 + 12
10	2 + 8	21	1 + 8 + 12

Figure 5.2. J. J. Thomson's electron ring arrangements of 1907.

published the first set of electronic arrangements (fig. 5.2), or what today we called electronic configurations. In taking this step Thomson went beyond Perrin and Nagaoka in suggesting that electrons went about moving in the atom in a structured manner.

Thomson based his ideas partly on the work of American physicist Alfred Mayer, who had experimented with magnets that he attached to corks and floated in a circular basin of water (fig. 5.3). Mayer had found that when up to five magnets were floated they would form a single ring, but that on the addition of a sixth magnet a new ring would be formed.[4] As more magnets were added the phenomenon was repeated, so that when a certain number of magnets was reached the addition of a new magnet caused the formation of yet another ring, thus producing an arrangement of concentric rings.

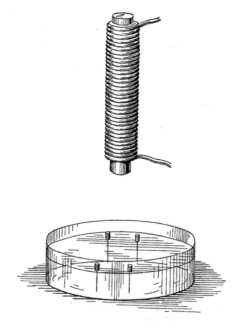

Figure 5.3. Alfred Mayer's magnets.

Thomson proposed that a similar principle might operate in the case of electrons circulating in the atom, and immediately began to consider how such a views might explain the periodic table in terms of electrons.

Thomson can therefore be regarded as the originator of electronic configurations and of attempts to explain the periodic table by means of such configurations. Fig. 5.2, reproduced from Thomson's article, shows how his electron rings were arranged. According to Thomson, and by analogy with Mayer's cork rings, the presence of five electrons results in the formation of one electron ring. A second ring begins to form once the number of electrons reaches six, although after this happens new electrons continue to be added to the first ring, again as in the case of the floating needles and corks. On reaching 10 electrons a new electron appears in the second ring, while on reaching 17 electrons a third ring begins to form. In each case the additional, or differentiating, electron is generally being added to an inner ring rather than an outer one.

From a modern point of view these electronic arrangements would seem to have little merit since they suggest a chemical analogy between element number 5, boron, and element number 16, sulfur, for example, which is not the case. But it would be a mistake to criticize Thomson on this point, since in 1904 nobody was aware of the number of electrons in any particular atom. In proposing his new scheme of electron rings Thompson was merely suggesting the plausibility of explaining periodicity through similarities in electronic structures among different elements, something which remains valid to this day.

Although Thompson's atomic model would soon be discarded by Rutherford, when he introduced his nuclear model of the atom,[5] it did succeed in establishing two important concepts. One was that the electron held the key to chemical periodicity, and the other was the notion that the atoms of successive elements in the periodic table differ by the addition of a single electron. Both of these ideas were to become important aspects of Bohr's atomic theory of periodicity, which would be published in 1913.

BOHR

In 1913 Niels Bohr published his spectacular quantum theory of the atom in a series of three papers that was subsequently named the "trilogy." One of these three parts was devoted to spelling out the electronic configurations of the atoms, supposedly on the basis of his newly developed quantum theory. There exists an overwhelming tendency in science, the history of science, and science education to believe that Bohr's arguments were solidly bolstered by his quantum theory. But as many observers commented at the time and many have done subsequently, it quickly became clear that Bohr was using a mixture of chemical spectroscopic arguments not to mention sheer trial and error.

> It could very distinctly be felt that Bohr had not reached his results through calculations and proofs but through empathy and inspiration and it was now difficult for him to defend them in front of the advanced school of mathematics in Göttingen.[6]
>
> After he had explained a simple spectrum he came to his crucial review of the structure of atoms with regard to their positions in the periodic system. In some respects this turned out to be obscure and not always easy to understand.[7]

In spite of Bohr's article there was still nothing like a detailed account of the configurations of all the atoms in the periodic table. Not surprisingly this task was better suited to a chemical approach, which is where Bury, Lewis, and Langmuir come into the picture.

KOSSELL

A good deal of Lewis's proposal, to be discussed in the next section, was arrived at simultaneously by the German physicist Walther Kossell working in Munich in 1916. Kossell had obtained his habilitation under Arnold Sommerfeld and had remained in Munich while working on his theory of bonding before accepting appointments at various other German universities.

Kossell's approach was directly centered on the notion that noble gas atoms are stable and unreactive, and that atoms placed either before or after noble gas atoms in the periodic table seem to aspire to this kind of stability. The way in which these other atoms can acquire such a status is either to lose one or more electrons, in the case of a metal, or to gain one or more electrons, in the case of a non-metal. The following are excerpts (translated from the original German) from Kossell's famous article of 1916.

> According to the van den Broek assumption, each successive element will contain one more electron and one more elementary quantity of positive charge than the previous one. This is primarily shown by the fact of the periodic changes of the valence number so that as the elements pass from lower to higher weight the configuration is not altered uniformly (also perhaps not if the newly arriving electrons are assumed to add on singly to the structure already formed arranged on a spiral). In the course of a regular change we are much more likely to come to a configuration in which the number of electrons capable of valence activity is repeated, and also some in which practically no tendency to exchange exists, the noble gases, among others.

Kossell proceeds with the suggestion that we focus on the beginning region of the periodic table, that is, from helium to titanium.

He notes that on adding eight electrons we obtain an atom with the same "surface order" or, as one might say today, the same outer-shell configuration. However, as he also explains, the observed frequencies of the K spectral lines, due to the inner electrons, vary continuously as we progress through the elements. This leads Kossell to conclude that the periodicity of the elements does not extend to the whole electron structure but only to its 'surface'.

> Accordingly, we can follow up the simplest idea that suggests itself: the order of the inner electrons remains unchanged in the elements that follow each other, always similar in each element and only changing its size through the continuous increase of the charge, which corresponds to the same increase in characteristic frequencies. The electrons that come anew in the sequence of atomic weights will always add to the outside and their order will be such that the observed periodicity results from the fact of their approach from outside.[8]
>
> In order to test these ideas more fully, we now ask in which column of the periodic table we must seek the elements in which a new shell from some electron lying outside is begun.

Kossell claims that this situation can be seen from strictly chemical considerations. For example, the characteristic of the alkali metals is to always give up electrons easily—which suggests that an electron is bound especially loosely and is exposed to outside influences. To Kossell this means that we can consider sodium and potassium as elements in which the first electron of the new shell is located outside the completed shell. Kossell proposes that we should "logically ascribe" a completely closed electron ring to the atoms of neon and argon. Meanwhile the elements preceding them, fluorine and chlorine, each lacks one electron in their outer rings and seek to acquire an additional electron. Kossell ascribes to them a "high affinity for electrons." He continues with,

It appears, then, that the configuration of the outer electrons reached in the noble gases can be considered somewhat analogous to an equilibrium state. Not only do the noble gases themselves lack an inclination, to take up electrons ("no affinity for free electrons") or to give them up ('they have the highest splitting tension yet measured), but also the neighboring configurations endeavor, by giving up or taking on electrons, to form systems of the same total number of electrons as the noble gases.

CONFIGURATIONS ACCORDING TO CHEMISTS

Lewis

As early as 1902 the chemist G. N. Lewis was already speculating about how many electrons would be found in each of the atoms of various elements and furthermore how these electrons were arranged. Lewis did not publish these early views because the prevailing academic climate in the United States was rather against theoretical speculations. He did nevertheless produce a manuscript in which he arranged electrons on the corners of cubes (fig. 5.4). The basis of this idea is the simple fact that, as we have seen, the periodic table involves an approximate repetition in properties after every eight elements.[9]

Writing about this work in later years Lewis said,

> In the year 1902 (while I was attempting to explain to an elementary class in chemistry some of the ideas involved in the periodic law) becoming interested in the new theory of the electron, and combining this idea with those implied in the periodic classification, I formed an idea of the inner structure of the atom which, although it contained certain crudities, I have ever since regarded as representing essentially the arrangement of electrons in the atom.[10]

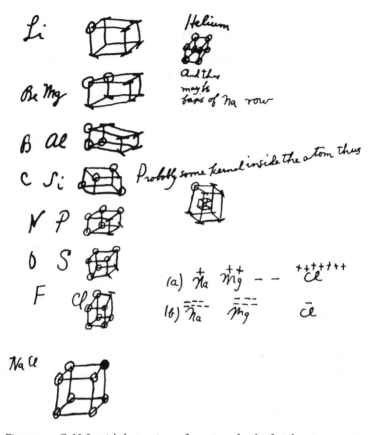

Figure 5.4. G. N. Lewis' electronic configurations for the first few atoms up to and including chlorine.

As is evident from Lewis's diagram, the periodicity in chemical properties is explained by the fact that elements falling in the same chemical group or column of the periodic table, such as lithium and sodium, have the same number of outer-cube electrons, namely one electron in these particular atoms. Similarly, Lewis regards beryllium and magnesium as possessing two electrons at two of the corners of concentric cubes and centered on the nucleus of their atoms. The model is a static one in which electrons seem to remain in fixed positions at the corners of cubes in contrast to the better known dynamical models of physicists such as Rutherford and Bohr for whom electrons circulate around the nucleus. Lewis would eventually abandon the view that electrons were static but as far as explaining chemical properties his original model is perfectly adequate.

Armed with these basic concepts, Lewis explained the formation of ionic or polar bonds, such as in the case of sodium chloride, as the transfer of one electron from the outer cube of the sodium atom to the empty space in the outer electron cube of the chlorine atom. This feature is illustrated by Lewis at the bottom left-hand corner of his diagram (fig. 5.4).[11]

It was not until 1916 that Lewis would publish these ideas in what has become a scientific classic entitled "The Atom and the Molecule."[12] Here he begins with the same idea of electrons at the eight corners of a cube and uses it to explain the other major kind of bonding than polar, namely non-polar or covalent bonds.[13]

In the same article, Lewis soon turns to an alternative, but still static view, of electrons arranged in pairs on the corners of a regular tetrahedron. An equivalent form of this arrangement consists of regarding the pair of electrons as lying at the center of four of the eight edges of a cube as shown in figure 5.5.

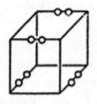

Figure 5.5. Lewis' pairs of electrons on four of the eight edges of a cube.

LANGMUIR

In 1919 the chemist Irving Langmuir, who spent most of his career at the General Electric Company, published a highly influential article entitled "The Arrangement of Electrons in Atoms and Molecules." In this extensive work (comprising 67 pages) he sets out to give improved electronic configurations as compared to those of Lewis and Kossell.[14] Langmuir begins his article with an interesting passage that highlights the tension between the chemical and physical approaches to the problem that was alluded to above:

> The problem of the structure of atoms has been attacked mainly by physicists who have given little consideration to the chemical properties, which must ultimately be explained by a theory of atomic structure. The vast store of knowledge of chemical properties and relationships, such as is summarized in the periodic table, should serve as a better foundation for a theory of atomic structure than the relatively meager experimental data along purely physical lines.

Langmuir goes on to mention the limitations of the electron arrangements published by his contemporary chemists and physicists, especially Lewis and Kossell. He points out that Lewis has only confined his attention to the main group elements and has not ventured to predict the arrangements of any transition metal atoms. He adds that Lewis has only managed to describe the arrangements in 35 of the then 88 known elements.

Meanwhile, Langmuir also points out that Kossell conceives of the electrons in dynamical arrangements but, instead of Lewis's cubes, supposes the electrons to be in the form of concentric rings rotating in orbits around the nucleus. Langmuir suggests four postulates to explain precisely how the electrons are arranged around the nucleus of any atom in a series of concentric shells. Such shells are further subdivided into what Langmuir terms cells in the following way (here we will cite just the second and third postulates, p. 870):

> Postulate 2. The electrons in the atom are distributed through a series of concentric shells. All the shells in a given atom are of equal thickness. If the mean of the inner and outer radii be considered to be the effective radius of the shell then the radii of the different shells stand in the ratio of $1 : 2 : 3 : 4$ and the effective surfaces of the shells are in the ratio of $1^2 : 2^2 : 3^2 : 4^2$.
>
> Postulate 3. Each spherical shell is divided into a number of cellular spaces. The thickness of these cells measured in the radial direction is equal to the thickness of the shell and is therefore the same (Postulate 2) for all the cells in the atom. In any given atom the cells occupy equal areas in their respective shells. All the cells in any atom have therefore equal volumes (fig. 5.6).

Turning to Kossell, Langmuir acknowledges that he provides arrangements for more elements than Lewis, namely the first 57 up to and including cerium. In doing so Kossell does consider some transition elements but, to Langmuir, his arrangements are "unsatisfying" for the elements vanadium to zinc in the first transition series; columbium[15] to silver in the second transition are also "unsatisfactory."

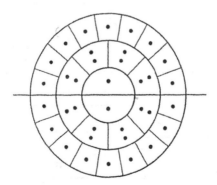

Figure 5.6. A representation of Langmuir's cells each occupied by up to two electrons.

Langmuir's periodic table of electronic arrangements (fig. 5.7) reveals some interesting idiosyncrasies, especially when viewed from a modern perspective. Among them are the appearance of what seem to be new forms of various elements such as nickel, palladium, erbium, and platinum. In all such cases Langmuir is postulating the existence of an alternative form of their atoms, which he designates as the β-form (fig. 5.8). In each case the β-form follows the regular or α-form of the atom in Langmuir's periodic table. A close inspection of Langmuir's table and the contents of his article shows that the β-form of nickel, to take the first and simplest example, consists of a slightly modified arrangement in the groups of five electrons at the upper and lower parts of the central cubic arrangement. Langmuir claims that the β-form possesses greater stability than the α-form

TABLE I.

Classification of the Elements According to the Arrangement of Their Electrons.

Layer.	N $E = 0$	1	2	3	4	5	6	7	8	9	10
I		H	He								
IIa	2	He	Li	Be	B	C	N	O	F	Ne	
IIb	10	Ne	Na	Mg	Al	Si	P	S	Cl	A	
IIIa	18	A	K	Ca	Sc	Ti	V	Cr	Mn	Fe	Co	Ni
			11	12	13	14	15	16	17	18		
IIIa	28	Niβ	Cu	Zn	Ga	Ge	As	Se	Br	Kr		
IIIb	36	Kr	Rb	Sr	Y	Zr	Cb	Mo	43	Ru	Rh	Pd
			11	12	13	14	15	16	17	18		
IIIb	46	Pdβ	Ag	Cd	In	Sn	Sb	Te	I	Xe		
IVa	54	Xe	Cs	Ba	La	Ce	Pr	Nd	61	Sa	Eu	Gd
			11	12	13	14	15	16	17	18		
IVa			Tb	Ho	Dy	Er	Tm	Tm$_2$	Yb	Lu		
			14	15	16	17	18	19	20	21	22	23	24
IVa	68	Erβ	Tmβ	Tm$_2\beta$	Ybβ	Luβ	Ta	W	75	Os	Ir	Pt
			25	26	27	28	29	30	31	32		
IVa	78	Ptβ	Au	Hg	Tl	Pb	Bi	RaF	85	Nt		
IVb	86	Nt	87	Ra	Ac	Th	Ux$_2$	U				

Figure 5.7. Langmuir's table of electronic arrangements.

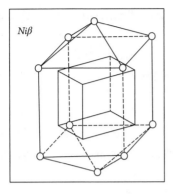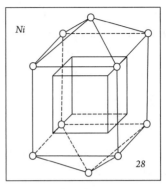

Figure 5.8. Langmuir's α and β forms of nickel.

because a rotation of the square pyramidlike structures at the top and bottom of the cube give rise to a more symmetrical arrangement.

Then comes a rather interesting move. Langmuir suggests that each atom close to nickel tries to achieve the stability of the β-form of nickel. To a contemporary reader the Langmuir suggestion would seem to be rather far-fetched if not downright bizarre. On the other hand this was just three years after Lewis and Kossell had suggested that main-group atoms attempt to reach the stability of noble gas atoms as the explanation for ionic bonding. Langmuir is merely extending this general notion in order to claim stability for certain elements like his β-nickel.

We are now finally in a position to discuss Charles Bury, the marginal scientific figure who is the main focus of the chapter.

A BRIEF BIOGRAPHY OF CHARLES BURY

Charles Bury was born in 1890 in the town of Henley-on-Thames, known as the site of the Henley regatta, an important date on the

English social calendar of the aristocracy. Bury's father was a solicitor who had graduated from Trinity Hall College at the University of Cambridge. The younger Bury attended the well-known public school, Malvern, and is said to have positively hated his time there. He then went to Trinity College, Oxford and obtained a first class honors degree; he was a contemporary of Henry Moseley, of atomic number fame.[16] After graduating from Oxford, Bury remained there for a while as a demonstrator in physical chemistry and conducted experiments on the conductivity of various salts and other solutions.

Bury then spent the academic year of 1912-13 at the University of Göttingen where he chose to work with the prominent physical chemist Walter Nernst. On returning to the UK he obtained an assistant lectureship at the University College of Wales in Aberystwyth, where he remained for the rest of his career. Very soon after taking up this post World War I broke out and like most young men did at the time, Bury volunteered to fight. Unlike Moseley though, Bury was to survive various major campaigns and returned to the UK in 1918 to pick up his experiments where he had left off.

It is, however, for a piece of theoretical work that Bury is known—to the extent that he is known at all, even to historians of science. As one of his students and admirers later wrote,[17]

> Rarely can it be true of a scientist's first paper, written without help or consultation with any colleagues, has such significance that all college students of the subject are necessarily taught its content. This however is true of Bury's first paper which bore the title: Langmuir's Theory of the Arrangement of Electrons in Atoms and Molecules.

Bury's paper provided nothing short of a detailed and successful explanation of the periodic table of the elements in terms of the electronic configurations of their atoms. This discovery is almost invariably attributed to Niels Bohr, the physicist who in fact receives the major share of the credit for most of the ideas discussed in this book. But, in fact, Bohr only gave a sketchy account of the configurations of atoms and moreover used chemical and spectroscopic evidence while frequently implying that he was deriving the configurations from the first principles of his quantum theory.[18]

Meanwhile Bury's name does not even appear in the 16-volume *Dictionary of Scientific Biography*.[19] Nor does his name appear in most histories of atomic physics. As one of Bury's students and protégées, Mansel Davies, wrote, "The simple fact is that anyone who learns the relationship between the electronic structures of chemical atoms and their placements in the Periodic table is absorbing an interpretation first arrived at and amply defined by Bury."[20]

It was left to the chemist, Neville Sidgwick, to provide the first extensive account of Bury's ideas on electron arrangements and their relationship to the periodic table in his well-known textbook written in 1927.[21] This book in turn was the source of information for many other textbooks that were subsequently written in the English language.[22]

BURY'S WORK

Bury begins his brief (7-page) article by launching into a critique of Langmuir's 1919 work on electronic arrangements of atoms. Bury especially objects to Langmuir's view that, "there can be no electrons in the outer shell until all the inner shells contain their maximum numbers of electrons."[23] Bury states from the outset that he intends to propose a set of different atomic structures that achieve several goals, including giving a better explanation of the chemical properties of the elements and avoiding certain inconsistencies in Langmuir's scheme, such as the haphazard assignment of one electron to some cells and two electrons to others.

But Bury prefers to speak of shells or layers rather than cells and suggests that the number of electrons in each shell should be proportional to their surface areas, namely 2, 8, 18, and 32 electrons. He then asserts that groups of 8 or 18 electrons are stable even if a particular shell can contain a larger number of electrons. This is a key departure from the view of previous authors who seemed committed to completely filling each shell or layer before needing to begin to fill a new shell. It is also a notion that persists in the current understanding of the periodic table and the electronic structure of transition as well as inner transition or f-block elements.[24]

Bury then makes a statement that that is no longer regarded as being correct.

> During the change of an inner layer from a stable group of 8 to one of 18, or 18 to 32, there occurs a transition series of elements which can have more than one structure.[25]

The modern view differs in that an atom has just one structure or one electronic configurations rather than several at the same time.

Bury agrees with Langmuir on the electron arrangements of what he calls the first two short periods.[26] He makes his disagreement with Langmuir clear when it comes to the third period: "Since 8 is the maximum number of electrons in the outer layer, potassium, calcium and scandium form a fourth layer, although their third is not complete. Their structures will be $(2,8,8,1)$, $(2,8,8,2)$ and $(2,8,8,3)$."

According to Bury the following is a list of electron arrangements for the elements titanium to copper (fig. 5.9):

Ti	(2,8,8,4)	(2,8,9,3)	(2,8,10,2)		
V	(2,8,8,5)	(2,8,9,4)	(2,8,10,3)	(2,8,11,2)	
Cr	(2,8,8,6)	(2,8,11,3)	(2,8,12,2)		
Mn	(2,8,8,7)	(2,8,9,6)	(2,8,11,4)	(2,8,12,3)	(2,8,13,2)
Fe	(2,8,10,6)	(2,8,12,4)	(2,8,13,3)	(2,8,14,2)	
Co	(2,8,13,4)	(2,8,14,3)	(2,8,15,2)		
Ni	(2,8,14,4)	(2,8,15,3)	(2,8,16,2)		
Cu	(2,8,17,2)	(2,8,18,1)			

Figure 5.9. Bury's alternative configurations for transition metals starting with titanium and ending with copper.

If one were to take a Whiggish view of this set of arrangements one might conclude that Bury was mistaken since it is now believed that atoms of each of the elements have just one electron arrangement rather than what is being proposed in this table.[27] On the other hand, it seems to provide yet another example of the gradually changing nature of scientific knowledge. Bury's view goes beyond Langmuir's idea, whereby new shells only start to fill after previous ones are completely filled but stops short of embracing the notion that atoms have a single arrangement of electrons. In fact, the notion of varying electron arrangements for the transition elements did not originate with Bury. It had previously been suggested by Saul Dushman, a colleague of Langmuir's at the General Electric Company.

Dushman was probably also the first author to try to connect the detailed chemical behavior of transition metals with electronic arrangements and their presumed variations. In his 1917 article Dushman presented the following arrangements for the element tungsten, or W, which he believed explain the fact that the element displays valences of 6, 5, 4, and 3, respectively:[28]

$$WO_3: 6\text{-valent}: 2, \quad 8, \quad 10, \quad 24, \quad 30 \quad [\text{i.e.}, 24 + 6]$$
$$WO_3: 5\text{-valent}: 2, \quad 8, \quad 10, \quad 25, \quad 29 \quad [\text{i.e.}, 24 + 5]$$
$$WO_3: 4\text{-valent}: 2, \quad 8, \quad 10, \quad 26, \quad 28 \quad [\text{i.e.}, 24 + 4]$$
$$WO_3: 3\text{-valent}: 2, \quad 8, \quad 10, \quad 28, \quad 26 \quad [\text{i.e.}, 24 + 3]$$

A fuller analysis of Dushman's work would provide an even finer grained examination of intermediate figures in the history of science than is being attempted in this book. This task will not be carried out here however.

Returning to Bury's article, after describing the rare earth elements he makes the prophetic remark, "Between lutecium and tantalum an element of atomic number 72 is to be expected. This would have the structure 2, 8, 18, 32, 8, 4 and would resemble zirconium."[29]

As Mansel Davies writes,

> Nothing could be clearer: element 72 would not be a rare earth but would be related to and (as is almost invariably true of other closely similar elements throughout the Periodic table) it would probably occur in association with zirconium.[30]

The importance of this part of Bury's work lies in the ensuing controversy over the discovery of element 72, which turned out to be one of the most bitterly fought scientific arguments in modern history. I have written a detailed version of this story in a couple of articles as well as in a recent book.[31] The brief version of this episode is that Niels Bohr is usually given the whole credit for having realized that element 72 would be a transition element rather than a rare earth. It is also usually said that it was Bohr who directed two of his researchers to look for this element in the ores of zirconium where it was indeed discovered.

The more accurate story is that Bohr was relying as much on chemical evidence as he was upon deductions from his quantum theory of atoms. Even more pertinent to the present case, Bohr acknowledged Bury's priority in the question of whether the element would be a rare earth or not.[32] In spite of this fact Bury has received very little credit for his priority and historians seem to have fallen prey to the reductionist tendency of favoring work done by physicists rather than chemists when it comes to atomic structure and the like. But whereas Bohr at least cited Bury's work, De Hevesy, one of the discoverers of element 72 (which he named hafnium), completely omitted any mention of Bury from his own writings.

One of the few authors to give ample recognition to Bury's work was, not surprisingly, a chemist—Samuel Glasstone, who even included a page heading with the words, "The Bohr-Bury" atom.[33] Finally, Bury makes another interesting prediction which has subsequently come true. Under the heading "The Last Period," Bury writes,

> In this period a second 18–32 transition series may be expected.... Little resemblance [between the actinides and lanthanides]... is to be expected Possibly an element, not yet discovered, of atomic number 94... is the first of a series of 7 transition elements... something like the ruthenium group but more electropositive.[34]

These comments are perhaps the first reasoned predictions of the existence of transuranic elements. Their history is a long and complicated one. Among other developments, the Italian physicist Enrico Fermi claimed that he had synthesized some of them in 1934. He was even awarded the Nobel Prize for this work but very soon afterward withdrew the claim. Meanwhile, the first genuine discovery of a transuranic element was made by Edwin McMillan in 1939.

CONCLUSION

Just as in the case of the other marginal figures discussed, the work of Bury reveals an assortment of ideas, some that survived and others that did not. Indeed this feature is common to all science and was seen in the brief excursion into the work of Langmuir on electronic configurations. While Langmuir's general ideas on transition metal configurations have been productive, his strange notions on two forms of nickel have sunk without a trace. This is just as one would expect on an evolutionary account of the development of science in which infertile notions fail to bear fruit and simply wither away while other aspects of a theory (other members of the species) live on and spawn yet new productive ideas (fitter offspring).

NOTES

1. There was even some reluctance expressed by Lewis, who was not keen to call it a rule.
2. The discovery is controversial, as most scientific discoveries tend to be.
3. Thomson himself never alluded to any plum pudding, as mentioned in an earlier chapter.
4. A.M. Meyer, *American Journal of Physics*, 116, 248, 1878.
5. Rutherford described Thomson's plum pudding model as, "old lumber fit only for a museum of scientific curiosities."
6. Heisenberg quoted in H. Kragh, "The Theory of the Periodic System," in A.P. French, P.J. Kennedy, eds., *Niels Bohr, A Centenary Volume*, Harvard University Press, Cambridge, MA, 1985, 50–67, quotation is from p. 61.
7. E. Hund, quoted in J. Mehra, H. Rechenberg, *The Discovery of Quantum Mechanics*, 1925, vol. 2 of *Historical Development of Quantum Theory*, Springer-Verlag, New York, 1982.
8. This changed in due course and the first to realize it was Charles Bury.
9. I am referring to the periodic table as displayed on an 8-column or so-called short form periodic table. The more generally used modern periodic table features an 18-column format but the importance of the number 8 is maintained even here. More specifically the closing of each period occurs at a noble gas element which possesses an outer shell of eight electrons.
10. G.N. Lewis, *Valence and the Structure of Atoms and Molecules*, Chemical Catalogue Company, New York, 1923, p. 29.
11. Quite remarkably, this simple explanation of how ionic bonds are formed, in terms of a transfer of electrons from metals to nonmetals was arrived at quite independently by the German physicist Walther Kossell as we have seen.
12. G.N. Lewis, "The Atom and the Molecule," *Journal of the American Chemical Society*, 38(4), 762–785, 1916.
13. In a recent article to celebrate the 100th anniversary of the landmark contributions by Lewis and Kosell, the inorganic chemist Michael Mingos makes the

following illuminating point: "What is remarkable is the success and wide-spread use of a model which stern critics would argue owes more to numerology than modern physics and was not solidly based on quantum or even Newtonian physics. In a contradictory manner, it (Lewis's model) defines the chemical bond in terms of a classical electrostatic interaction between oppositely charged ions (the ionic bond) and the pairing of negatively charged electrons sharing a small region of the molecule (the covalent bond). It is hardly surprising that the contradiction made Lewis delay publication from 1902 when he first introduced the basic ideas to undergraduates in his lectures." M. Mingos, in *The Chemical Bond II, Structure & Bonding* series vol. 170, Springer, Berlin, 2016.

14. W. Kossel, "Molecule Formation as a Question of Atomic Structure," *Annalen der Physik*, 49, 229–362, 1916.

15. Columbium is an older name for element 91 in the periodic table, now called niobium.

16. It is not known whether the two even knew or met each other. There is also no correspondence between the two men in Heilbron's list of Moseley's letters.

17. M. Davies, "Charles Rugeley Bury and his Contributions to Physical Chemistry," *Archive for the History of the Exact Sciences*, 36, 75–90, 1986, quotation is from p. 78.

18. I have argued this point in greater detail elsewhere and will not repeat myself here. E.R. Scerri, *The Periodic Table, Its Story and Its Significance*, Oxford University Press, New York, 2007.

19. *Dictionary of Scientific Biography*, editor-in-chief, C.C. Gillespie, Charles Scribner's, New York, 1970–1976.

20. M. Davis, "Charles Rugeley Bury and his Contributions to Physical Chemistry," *Archive for History of Exact Sciences*, 36, 75–90, 1986, quotation is from p. 75

21. N.V. Sidgwick, *The Electronic Theory of Valency*, Clarendon Press, Oxford, 1927.

22. K. Gavroglu, A. Simoes, "Preparing the Ground for Quantum Chemistry in Great Britain: The Work of the Physicist R.H. Fowler and the Chemist N.V. Sidgwick," *British Journal for the History of Science*, 35(2), 187–212, 2002.

23. I. Langmuir, "The Arrangement of Electrons in Atoms and Molecules" *Journal of the American Chemical Society* 41 (6): 868–934, 1919, quotation is from p. 871.

24. Bury can be said to have begun the now very well-trodden discussion on the relative occupation of the 4s and 3d orbitals. Some more recent articles on this theme are; F. Pilar, "4s Is Always above 3d!," *Journal of Chemical Education*, 55, 2–6, 1978; L.G. Vanquickenborne, K. Pierloot, D. Devoghel, "Transition Metals and the Aufbau Principle," *Journal of Chemical Education*, 71, 469–471, 1994; E.R. Scerri, "Transition Metal Configurations and Limitations of the Orbital Approximation," *Journal of Chemical Education*, 66, 481–483, 1989; M.P. Melrose, E.R. Scerri, "Why the 4s Orbital Is Occupied Before 3d," *Journal of Chemical Education*, 73, 498–503, 1996; W.H.E. Schwarz, "The Full Story of the

Electron Configurations of the Transition Elements," *Journal of Chemical Education*, 87, 444–448, 2010; E.R. Scerri, "The Trouble with the Aufbau Principle," *Education in Chemistry*, November, 24–26, 2013.

25. C.R. Bury, "Langmuir's Theory on the Arrangement of Electrons in Atoms and Molecules," *Journal of the American Chemical Society*, 43, 1602–1609, 1921, quotation is from p. 1602.

26. What is meant here is elements lithium to argon. The very short 2-element period consisting of hydrogen and helium is not being counted.

27. As mentioned in chapter 1, modern work also goes beyond single configurations but in a rather different sense of admitting the superposition of numerous electronic configurations. This is not what Bury was referring to, however, in writing several configurations for some transition metals.

28. S. Dushman, "The Structure of the Atom," *General Electric Review*, 20, 186–196, 397–411, 1917.

29. C.R. Bury, "Langmuir's Theory on the Arrangement of Electrons in Atoms and Molecules," *Journal of the American Chemical Society*, 43, 1602–1609, 1921.

30. M. Davis, "Charles Rugeley Bury and His Contributions to Physical Chemistry," *Archive for History of Exact Sciences*, 36, 75–90, 1986, quotation is from p. 80.

31. E.R. Scerri, "Prediction of the Nature of Hafnium from Chemistry, Bohr's Theory and Quantum Theory," *Annals of Science*, 51, 137–150, 1994; E.R. Scerri, *A Tale of Seven Elements*, Oxford University Press, New York, 2013, chapter 4.

32. N. Bohr, "The Structure of the Atom," *Nature*, 112, 29–44, 1923.

33. S. Glasstone, *Recent Advances in Physical Chemistry*, J. & A. Churchill, London, 1931.

34. C.R. Bury, "Langmuir's Theory on the Arrangement of Electrons in Atoms and Molecules," *Journal of the American Chemical Society*, 43, 1602–1609, 1921, quotation is from p. 1606.

Chapter 6

John D. Main Smith

Very little is known about John D. Main Smith, the next personality to be examined (fig. 6.1). For example, I have only been able to locate a single photograph of him.[1] What is known about his life is that he was a chemist at the University of Birmingham in the UK who wrote a comprehensive book entitled *Chemistry & Atomic Structure* in 1924. His small claim to fame, which should be much greater than it is, rests mainly with a handful of articles that he published in the years 1924–25, in addition to his book. Unfortunately Main Smith's articles appeared in a rather obscure journal, *Chemistry & Industry*, with the result that, with a few minor exceptions, they did not have any serious influence on chemists and certainly not on physicists.

As we saw in the previous chapter, the chemist Charles Bury took on the work of Irving Langmuir and subjected it to detailed criticism on the basis of a careful analysis of the chemical behavior of all the elements (rather than just some of them). As we also saw, Bury was successful in obtaining a better and more complete set of electron arrangements which provided an improved account of the periodic table of the elements.

On the other hand, three years later, the chemist Main Smith took on not just his fellow chemists but the mighty physicist, Niels Bohr, and succeeded in proposing some improved electronic arrangements. These new arrangements were subsequently and independently rediscovered by a young English theoretical physicist,

Figure 6.1. John Main Smith.

Edmund Stoner (who will be examined in the following chapter).[2]

Main Smith's critique included the claim that the views of Lewis and Langmuir have no mathematical or dynamic necessity consequent on the electron configurations of the suggested Bohr atoms, but are extraneous assumptions arising from the absence in Bohr's theory of any provision for the combination between atoms without ionization.[3]

In contrast to Lewis' famous notion of a covalent bond as consisting of a pair of electrons, Main Smith held the view that such a bond consisted of just one electron and he cited the case of the H_2^+ ion as an example of his view.[4] Whereas Bury's approach is almost exclusively based on chemistry and is devoid of a quantum theoretical analysis, that of Main Smith shows a deep understanding of quantum physics. It might therefore not be an exaggeration to suggest that Main Smith took the fight to the physicist and won it—barring what I say about winners and losers in science, of course. We have already touched on the reasons his contribution has been almost obliterated from all but specialist histories of physics, but we will need to examine the work more closely in order to see just what he achieved.

Main Smith's contribution may be put very simply by saying that he challenged Bohr's view of a symmetrical distribution of electrons in each shell surrounding the nucleus of an atom. For example, in the case of the second shell, which had long been known to contain eight electrons, Bohr thought the structure to consist of two subshells each containing four electrons. By means of a detailed analysis of chemical evidence, the unknown chemist Main Smith had the temerity to challenge this view and to claim that the second shell should be regarded as having a grouping of 2, 2, and 4 electrons. Similarly Main Smith held that every shell begins with a subshell containing just two electrons. In contemporary terms he could be said to have discovered what is now termed an s-orbital, meaning a subshell that can hold a maximum of just two electrons—and the fact that each shell begins with such an s-orbital.

There had already been some published criticisms of Bohr's electron arrangements, and more specifically those by the French pair De Broglie and Dauvillier. In a couple of 1924 articles these authors analyzed the intensities in the X-ray spectra of several elements and found discrepancies with the assumptions made by Bohr.[5] However, they had not proposed any modified electron arrangements.

Meanwhile Main Smith went a good deal further. A brief passage from his book serves to illustrate the depth of his disagreement with Bohr:

It will be shown in the next chapter that Bohr's subgroups do not fit chemical facts. It inevitably gives symmetrical arrangements because the total number of electrons in a group is twice the square of the total quantum number, i.e. Rydberg's series, the unsquared numbers of which are accidentally the same as Bohr's total quantum numbers.... It is unfortunate that Bohr was misled by a numerical coincidence and the properties of numbers, because his subgroup scheme has been made the basis of theories of chemical combination which come near enough to agreeing with the chemical facts to make them a valid expedient in chemistry, and have led to a non-concordant interpretation of chemical problems of atomic and molecular structure and the mechanism of electronic combination.[6]

The following passage is taken from the final chapter of Main Smith's book.

> Bohr's theory of atomic structure is strictly a theory relating to single atoms, neutral or ionized, far removed from the influence of other atoms. The fact that it is an interpretation of the periodic classification of the elements, largely based on the properties of atoms in combination, indicates that it must be valid for atoms in combination, at least so far as the broad outlines of the theory are concerned. The theory certainly necessitates that the general type of structure of an atom is preserved even when the atom loses some of its electrons and thus acquires positive charges.

> A little later he writes
>
> This is further confirmed by Moseley's spectroscopic law connecting the wavelength of X-irradiation with atomic number and with quantum conditions of atoms, the various quantum conditions being unchanged in atoms in combination. It may therefore be accepted that the structure of an atom is not influenced in qualitative aspects and only slightly in quantitative aspects by chemical combination. In general it must be assumed that chemical combination between atoms is powerless to alter more than the external portions of the structure of atoms and that this alteration occurs in or near the electrons of the structure which are known as valence electrons.

Main Smith reasons that one of the outstanding periodic properties of metal atoms in general is the tendency to yield compounds which are alkaline, acid, and amphoterically neutral. He adds that an examination of the periodic system indicates that the property of alkalinity is intense for the univalent alkali metals and is less so for the bivalent alkali metals. Alkalinity suddenly disappears in trivalent elements boron, aluminium, gallium, and so on, and does not appear elsewhere in the periodic classification. Meanwhile, increasing acidity is observed as the periods are traversed. Main Smith remarks on the property of the trivalent elements of group III to yield compounds containing two valences, one coordinated or non-ionizing, and one un-coordinated or ionizing valence. The property of the two coordinated valences is common to the whole of the quadrivalent elements of group IV, and persists to a considerable extent in the elements of groups V, VI, and VII. This property is only observable when the elements are acting electopositively, that is, when oxidized or partly deprived of electrons. Main Smith writes that this may be interpreted as evidence that all elements containing more than two valence electrons have two electrons more firmly attached than other valence electrons. He interprets this further in terms of orbits as indicating that two electrons in the outer structure of atoms are in quantum orbits, the energies of which are different from that of out of other outer electrons.

Returning to Main Smith's own words,

This is directly confirmed by the well-known spectroscopic fact that the outer orbits of the aluminium atom with three valency electrons of two different energy types, and of the spectrum of trivalent elements generally bear a considerable resemblance to the spectrum of univalent elements. Corresponding observations on the property of acidity of elements indicate that it is a maximum in group VII and diminishes only gradually in groups VI and V and suddenly changes to perceptible basicity, or at least amphotericity in group IV. It is further notable that nearly all of the elements of these groups have marked tendency to yield compounds having

TABLE 12
GENERAL SCHEME OF ATOMIC STRUCTURE

Total quantum number $n=$				1	2	3	4	5	6
Azimuthal quantum number $k=$				1	112	11223	1122334	11223	1122
He (2)	-	-	-	2					
Gl (4)	-	-	-	2	2				
C (6)	-	-	-	2	22				
Ne (10)	-	-	-	2	224				
Ar (18) and Sc³ (21)		-		2	224	224			
Cu¹ (29)	-	-	-	2	224	22446			
Kr (36) and Y³ (39)		-		2	224	22446	224		
Ag¹ (47)	-	-	-	2	224	22446	22446		
Xe (54) and La³ (57)		-		2	224	22446	22446	224	
Lu³ (71)	-	-	-	2	224	22446	2244668	224	
Au¹ (79)	-	-	-	2	224	22446	2244668	22446	
Nt (86)	-	-	-	2	224	22446	2244668	22446	224
U² (92)	-	-	-	2	224	22446	2244668	22446	2244

TABLE 13
TRANSITION SERIES I

Total quantum number $n=$	1	2	3							
Azimuthal quantum number $k=$	1	112	112	23	23	23	23	23	23	23
Valency of ion	—	—	—	1	2	3	4	5	6	7
Sc (21) - -	2	224	224	—	—	00				
Ti (22) - -	2	224	224	—	02	01	00			
V (23) - -	2	224	224	—	03	02	01	00		
Cr (24) - -	2	224	224	—	04	03	—	01	00	
Mn (25) - -	2	224	224	—	14	13	12	—	10	00
Fe (26) - -	2	224	224	—	24	23	—	—	20	
Co (27) - -	2	224	224	—	25	24	23			
Ni (28) - -	2	224	224	—	26	—	24			
Cu (29) - -	2	224	224	46	45	44				

Figure 6.2. Tables of electron configurations from pp. 196–198 in Main Smith's book.

the coordination number 4, independently of the number of ions with which the coordination complex maybe associated. This may be interpreted as evidence that the valence electrons in excess of four are all equally feebly attached to the atoms, which further interpreted in terms of orbits, indicates that the valence electrons in excess of four are all in similar quantum orbits the energy of which is less than that of the first four electrons. The detailed chemical evidence by which it can be shown that the first two valence electrons are differentiated in energy or firmness of binding from the third and fourth electrons, is so vast that no more than a tithe of this evidence need to cited to prove the point.[7]

A portion of Main Smith's detailed electron arrangements can be seen in figure 6.2, which is also taken from his book.

SOME CURIOUS ASPECTS OF MAIN SMITH'S WORK

A major theme of this book is that scientists are neither right nor wrong—or said another way, it does not matter whether they are right or wrong provided that at least something that they contributed is accepted into the collective store of scientific knowledge, or that their views serve to catalyze those of others.

In his classic article of 1927 Main Smith provides some interesting examples of apparently being completely wrong on a specific issue having to do with the elements in the eighth period, namely the period starting with element 87, francium. Main Smith writes, "The chemical and physical properties of actinium, thorium, protactinium and uranium indicate decisively that they are analogues of yttrium, zirconium, etc. and therefore transition elements."[8]

A few lines later the view is reinforced further:

The existence of only one element, actinium, between thorium and radium renders it certain that the transition series of this last long period contains no transition sub-series similar to the analogous 14 "rare earth" elements, and it is consequently probable that, if all the elements of this period were known, they would number only 18 as in the case of the first and second periods.[9]

However, this view stands in complete opposition to the currently accepted wisdom whereby there is indeed an analogous long period of 32 elements and that starting with actinium, we do see a series of elements that are analogous to the rare earth elements and which in Main Smith's words could be said to constitute a transition subseries. This modern view is generally attributed to the American chemist Glen Seaborg, who proposed his amendment to the periodic table in the 1940s. As we will see in chapter 8, Charles Janet, one of the seven lesser-known scientists that are featured in this book, unambiguously anticipated this idea.

Returning to Main Smith, there is an interesting discussion of the placement of hydrogen in the periodic system where he expresses a view that has been receiving a good deal of attention in recent years.[10]

As the unit of valency and of atomic weight, hydrogen may be regarded as the type of all valent elements and placed at the head of any of the groups I to VII of the classification, helium being placed at the head of group VIII as the type of all the non-valent and noble elements. It is not a matter of importance, in the classification of the elements, in which group hydrogen is placed, but from its uni-valency it may conveniently be placed in Group I *and* alternatively in Group VII.[11]

Here is another example of the mixed nature of Main Smith's work as judged by a reviewer of his book:

Dr. Main Smith's book is not easy to review, in part because of the curious mixture of skepticism and speculation with which its more solid qualities are associated. This mixture is not unusual, since a skeptic in reference to orthodox views is often credulous in reference to heterodox alternatives. In the present instance, the skepticism is mainly in reference to the ability of modern physics to cover with their theories the facts which are familiar to chemists; whilst the speculations take the form of a chemist's guesses in regions in which all the experimental data have been supplied by physicists. Thus on the one hand the author lays much stress on "the incorrectness of Bohr's theory of the detailed structures of electron sub-groups", and asserts rather dogmatically (on the authority of Professors Thorpe and Morgan) that "the time is not yet ripe for the application of general electronic theories to organic chemistry"; on the other hand he advances, as it were a plausible alternative, the view that it may be the electric charge of a particle and not its mass that varies with its velocity, and puts forward the suggestion that, in Rutherford's experiments on the artificial disintegration of atoms, it may be the α-particle that provides the hydrogen and not the bombarded atoms.[12]

THE INERT PAIR EFFECT

Another concept that can definitely be attributed to Main Smith as well as to Stoner is one that was named the inert pair effect by the chemist Neville Sidgwick. The essential idea is that as one descends several groups in the p-block of the periodic table there is an increasing tendency for two of the outermost electrons to not be involved in covalent bonding. For example the elements tin and lead in group 14 of the periodic table form di-chlorides whereas the higher members of the group such as carbon and silicon invariably form tetra-chlorides. The following passage reveals that the origins of the concept belong with Main Smith and Stoner, both of whom stressed the importance of regarding two of the outer electrons as being grouped together, in contrast to Bohr who regarded a shell of eight electrons, for example, to consist of two groups of four electrons.

There is however one considerable class of abnormal valencies which can be recognized as being due to some common cause, that, namely, in which two of the valency electrons appear to have become inert, as though they had been absorbed into the core. This phenomenon is first to be noticed in IIIB[13] in InCl, and more remarkably in the thallous compounds: in the typical and B elements of group IV it is especially prominent, from the divalent compounds of carbon, to those of germanium, tin and lead. In group VB we find it in antimony and still more in bismuth, in VIB in sulphur, selenium and tellurium, and in group VIIB in iodine and bromine and possibly in chlorine. Thus it is most marked in the fourth group and is generally prevalent among the heavier B elements.... The new development of the Bohr theory due to Stoner and Main Smith indicates a possible reason for this peculiarity. We now realize that the first two electrons in any group ... correspond to the pair of electrons in helium, and can have a certain completeness of their own, approaching that which they have in helium, where they form a complete group.[14]

The earlier explanation of the inert pair effect suggested by Sidgwick, and others, was that the valence electrons in an s-orbital are more tightly bound and are of lower energy than electrons in p-orbitals and

Ionization Potential	Boron	Aluminum	Gallium	Indium	Thallium
2nd	2427.1	1816.7	1979.3	1820.6	1971
3rd	3659.7	2744.8	2963	2704	2878
(2nd + 3rd)	6086.8	4561.5	4942.3	4524.6	4849

Figure 6.3. Ionization potentials for group 13 elements in kJ/mol.

therefore less likely to be involved in bonding. However this explanation was subsequently found wanting. If the total ionization potentials of the 2 electrons in s-orbitals, that is the 2nd and 3rd ionization potentials, are examined it can be seen that they increase in the sequence, In < Al < Tl < Ga, which does not correlate with the fact that the inert pair effect becomes more and more pronounced as one descends the group (fig. 6.3).

Various authors, starting with Drago in the 1950s, have examined this situation in detail.[15] More recently, the higher sum of the 2nd and 3rd ionization potentials of thallium relative to indium, has been explained as a relativistic effect.[16]

These complications have resulted in the notion that the term inert pair effect should be viewed as a description of the phenomenon rather than as its explanation.[17]

CONCLUSION

Finally, Main Smith was well aware of the neglect of his work and tried to correct the situation by submitting the following letter to the *Philosophical Magazine*. This document is important since he is claiming priority while at the same time acknowledging the fact that Stoner arrived at the same conclusions in an independent manner.

Gentlemen, -

I shall be glad if you will allow me to direct attention to the fact that the distribution of electrons in atoms characterized by the sub-grouping 2: 2, 2, 4, 4, 6: 2, 2, 4, 4, 6, 6, 8, did not originate with Mr. E.C. Stoner, as within recent months several papers published in your magazine have suggested. Mr. E. C. Stoner's papers suggesting this distribution of electrons appeared in your issue published on 1st October, 1924. This electronic distribution was with all relevant detail, however, originally proposed by me and was published six months earlier in the issue of 28th March, 1924, of the "Review of Chemistry and Industry," and reprinted with the chemical evidence proving the exact distribution of the electrons in every known element in terms of this scheme, and elaborated in detail in my book "Chemistry and Atomic Structure" published in October 1924. I do not suggest that Mr. E.C. Stoner's work was not independent, but in view of the foregoing there can be no question of my priority in the matter of this electronic distribution of electrons in atoms. On the Continent and in America my priority is acknowledged, but in my own country my work has failed even to be cited in papers in your magazine.[18]

To maintain my evolutionary analogy I suggest that Main Smith might be regarded as one of the missing links in the history of atomic science. Main Smith would therefore be like one of those alleged missing links that the Creationists so love to focus upon, except that generally speaking they are not missing in evolutionary biology either.[19] As we will see in a later chapter concerning a contemporary of Main Smith's, the physicist Edmund Stoner independently arrived at many of the same key ideas as Main Smith. He too is a missing link, although he appears to be far more visible in historical accounts that the almost completely unknown chemist from Birmingham, John Main Smith.

NOTES

1. This was only achieved when I eventually tracked down his son, Bruce Main Smith, with the help of Peter Morris. None of the institutions that Main Smith was associated with nor any science libraries have any photographs of him.
2. Main Smith complained about the fact that his prior discovery had been overlooked although he acknowledged the fact that Stoner had arrived at the same electron arrangements independently.
3. J. Main Smith, "The Bohr Atom," *Chemistry and Industry*, 42, 1073–1078, 1923, quotation is from p. 1073.
4. This is the only molecule for which the Schrödinger equation possesses an exact, analytical, solution.
5. L. de Broglie, A. Dauvillier, "Le système spectrale des Rayons Röntgen et la structure de l'atome," *Journale de Physique*, 5, 1–9, 1924; A. Dauvillier, "Sur la distribution des electrons dans les niveaux L des éléments," *Comptes Rendus de l'Academie des Sciences*, 178, 476–479, 1924.
6. J. Main Smith, *Chemistry and Atomic Structure*, Ernest Benn, London, 1924, p. 186.
7. Ibid., pp. 189–191.
8. J.D. Main Smith, "The Electronic Structure of Atoms. Part I. The Periodic Classification," *Journal of the Chemical Society*, 1927, 2029–2038, quotation is from p. 2032.
9. Ibid., p. 2032.
10. See an extensive discussion of the placement of hydrogen and whether there exists an optimal periodic system in chapter 10 of E.R. Scerri, *A Very Short Introduction to the Periodic Table*, Oxford University Press, Oxford, 2011.
11. Ibid., p. 2033.
12. T.M. Lowry, "Review of Main Smith's, Chemistry and Atomic Structure," *Chemistry and Industry*, March 13, 287–287, 1925.
13. The use of letters A and B to identify groups of the periodic table is now obsolete. In Britain, where Sidgwick was writing, the B groups denoted the p-block elements whereas the A label referred to the transition metal groups. In the US the opposite convention was used. The confusion that this caused contributed to the eventual rationalization recommended by IUPAC (International Union of Pure and Applied Chemistry) whereby a numbering system of groups 1 to 18 was introduced and the A and B labels were abandoned.
14. N.V. Sidgwick, *The Electronic Theory of Valency*, Oxford University Press, Oxford, 1927, quotation is from pp. 178–179.

15.   Russell S. Drago, "Thermodynamic Evaluation of the Inert Pair Effect," *Journal of Physical Chemistry*, 62, 353–357, 1958.

16.   A.F. Holleman, E. Wiberg, *Inorganic Chemistry*, Academic Press, San Diego, 2001.

17.   Greenwood, Norman N., *Earnshaw, Alan, Chemistry of the Elements*, 2nd ed., Butterworth-Heinemann, 1997.

18.   J. Main Smith, "The Distribution of Electrons in Atoms [letter dated September 8]," *Philosophical Magazine*, 50 (6), 878–879, 1925, quotation is from p. 878.

19.   D.R. Prothero, *Evolution: What the Fossils Say and Why It Matters*, Columbia University Press, New York, 2007.

Chapter 7

Edmund Stoner

Edmund Clifton Stoner (fig. 7.1) was born in 1899 in Esher, in the county of Surrey in Southeast England. His father died a few months before his birth. Stoner's father had wanted to be a schoolteacher but financial hardship had forced him to leave school and eventually becoming a professional cricketer. His son, Edmund Stoner received an education that depended entirely on his ability to win scholarships, something in which he excelled; for example, he earned a scholarship to study at the Bolton Grammar school in the 1910s. A few years later and unlike Henry Moseley, Stoner was spared the horrors of the First World War due to his poor health.

In 1918 Stoner was awarded a scholarship to study at the University of Cambridge where he is known to have endured financial hardship. He later complained of the lack of mathematical training given to physicists and how "absurd" it was that he should have ended up as theoretical physicist.

Around 1919 Stoner began to suffer from bad health and after reading a number of medical textbooks he reached the correct conclusion that he was suffering from diabetes. Unfortunately, this happened before the discovery that insulin could be used to treat diabetes (following the work of Banting and Best in 1921); however Stoner was able to carefully control his dietary intake of carbohydrates, thereby experiencing a substantial improvement in his health. He began to benefit from the use of insulin in 1923 after being admitted to Addenbrookes Hospital in Cambridge following a bout of illness.

Figure 7.1. Edmund Stoner.

But it was not until 1927 that he was able to get a long-term prescription to use insulin in order to treat his diabetic condition effectively.

Throughout this period Stoner was a researcher at the Cavendish laboratory in Cambridge, with Ernest Rutherford as his supervisor. It was about this time that Stoner submitted a research proposal to work on X-ray absorption, a proposal which Rutherford approved and for which he arranged the purchase of the necessary apparatus.

Stoner later wrote the following about Rutherford:

> When things were going badly however, he [Rutherford] could make the most devastating comments in his naturally loud voice which could be heard far away.... I could never accustom myself to Rutherford's "bark", not to his forceful dominance in discussions; and except once after my research period, when I was talking to him in his own home, I never found conversation with him easy.[1]

In contrast to experimental research, which Stoner found boring, he was thrilled and excited by Bohr's theory of the hydrogen atom. In March of 1922 he attended Bohr's lectures in Cambridge, on this subject, and wrote to his mother,

> Quantum Theory is absolutely at variance with the previous scientific views (and apparently irreconcilable with them) & yet, when applied to the atom, leads to results which are borne out by experiment with extraordinary accuracy.... Such a state of affairs, as you can imagine, is very exciting.[2]

In 1924 Stoner undertook the work for which he is known among historians of physics, although generally not by the physics community. He became very familiar with the nomenclature in X-ray and optical spectroscopy as well as with their quantum specifications (as he put it). He devoted a great deal of attention to the question of the distribution of electrons among these levels, culminating in a report that he deposited on Rutherford's desk. Rutherford in turn passed it on to R. H. Fowler who was impressed by its contents and urged Stoner to write a full article that soon appeared in the *Philosophical Magazine* under the title of "On the Distribution of Electrons Among Atomic Levels."

AN INTERLUDE ON BOHR'S THEORY OF THE PERIODIC TABLE

Before describing Edmund Stoner's work further, it is necessary to understand the context of his discovery by delving into Bohr's theory of electronic configurations. Bohr's model of the hydrogen atom appeared to provide a solution to two deep problems that existed in the physics of his era. First, Rutherford's model of the atom suffered from a fundamental instability. According to classical electromagnetic theory, and more specifically an equation due to Larmor, any electron performing an acceleration would be subject to the rapid loss of energy, leading to its collapsing into the nucleus of the atom, something that of course does not take place. Moreover, Rutherford's model gave no hint as to why the spectrum of the hydrogen atom, or indeed any atom, gave a series of discrete lines.

The notion of quanta, or packets of energy, had been introduced by Max Planck in 1900 to explain the details of observations made on the spectrum of black-body radiation.[3] Bohr adopted Planck's notion of quantization and applied it to the physics of atoms, following some early hints provided by Nicholson (as we saw in a previous chapter).

What is not so well known is that Bohr's 1913 theory of the hydrogen atom also provided the first successful explanation of the periodic system in terms of arrangements of electrons in the atom. In addition Bohr's theory was aimed primarily as a criticism of an earlier model of the atom than Rutherford's, namely that of J. J. Thomson.

Bohr's calculations led him to conclude that, among other features, additional rings of electrons formed outside already full rings, thus correcting the earlier Thomson model, in which electrons are added to inner rings. In a somewhat ad hoc manner, Bohr proposed that electrons would be stable if they remained in certain quantized orbits and would lose energy only on undergoing transitions from one orbit to another, more stable orbit.[4] Electrons in a discrete set of stable orbits around the nucleus of an atom were regarded by Bohr as being in stationary states that would not radiate energy.[5]

Bohr was thus following Planck's quantum program in departing from classical electromagnetic theory. Just as Planck had found it necessary to introduce a constant, h, the elementary quantum of action, to explain its discontinuous nature, Bohr did much the same regarding the energy, and a little later, the angular momentum of orbiting electrons. Bohr essentially proposed that for the atom to pass from one energy state to another it must either emit or absorb one quantum of energy.

Unfortunately Bohr's theory was limited in its application, giving an exact prediction for only the spectrum of the hydrogen atom. Atoms with more than one electron undergo a complicated set of interelectronic repulsions among each other that Bohr did not have the resources to calculate precisely. All the same, Bohr had sufficient confidence in his quantum theory of the atom to apply it to multielectron atoms in an approximate manner.

As Heilbron and Kuhn have argued, Bohr's attempt to understand the periodic table and to examine the stability of Thomson's electron rings came first. Bohr's better known assault on the problem of the instability of Rutherford's atom came later, although the latter became the opening issue in Bohr's three-part trilogy of 1913.[6]

In the same series of three article published in the *Philosophical Magazine*, Bohr produced his first version of what may be termed an electronic periodic table (fig. 7.2). In it he assigned electronic configurations to the atoms of various elements in terms of the principal quantum number of each electron, which he used to characterize the stationary, or nonradiating, states.[7]

He called his method the *aufbauprinzip*, and it comprised building up atoms of successive elements in the periodic table by the addition of an outer electron to the previous atom, although there were exceptions to this rule, as we will discuss later. At specific stages in this process, a shell would become full, at which point a new shell would begin to fill. Contrary to the impression that he created in his published articles, however, Bohr was unable to deduce the maximum capacity of each electron shell, and he was guided entirely by chemical and spectroscopic data rather than by theoretical derivation.

For example, Bohr determined the number of electrons in the outermost ring of the atom of each element by considering its chemical valence. In the case of the nitrogen atom, which possesses seven electrons, Bohr was obliged to rearrange the inner shell in order to produce a configuration that corresponds to the element's known trivalence, as can be seen in figure 7.2. Whereas Bohr's building-up of atoms from hydrogen to carbon held that atoms have two inner electrons and a varying number of outer electrons, once nitrogen is reached, the inner electron shell abruptly doubles. This move appears to be made precisely so that Bohr can obtain the three outer electrons needed to correspond to the nitrogen atom's formation of three

1	H	1				
2	He	2				
3	Li	2	1			
4	Be	2	2			
5	B	2	3			
6	C	2	4			
7	N	2	3			
8	O	4	2	2		
9	F	4	4	1		
10	Ne	4	2			
11	Na	8	2	1		
12	Mg	8	2	2		
13	Al	8	2	3		
14	Si	8	2	4		
15	P	8	4	3		
16	S	8	4	2	2	
17	Cl	8	4	4	1	
18	Ar	8	8	2		
19	K	8	8	2	1	
20	Ca	8	8	2	2	
21	Sc	8	8	2	3	
22	Ti	8	8	2	4	
23	V	8	8	4	3	
24	Cr	8	8	2	2	2

Figure 7.2. Bohr's 1913 scheme for electronic configurations of atoms showing number of electrons in consecutive levels, beginning closest to the nucleus.

chemical bonds as in the case of ammonia (or NH_3)—in other words, the move is ad hoc.

In any case, Bohr gave no theoretical arguments for why such a rearrangement should occur, and such abrupt rearrangements could be seen in a number of places just among the 24 configurations shown in figure 7.2. Instead of rigorously deriving electronic arrangements from quantum theory, Bohr was leaning on intuition as well as spectroscopic and chemical behavior.

Nevertheless, Bohr achieved at least two goals with his semi-empirical approach. First he introduced the idea that the differentiating electron should, in most cases, occupy an outer shell and not an inner one, as Thomson had supposed. Second, in spite of its semi-empirical aspects, Bohr's scheme provided a correlation between electronic configurations and chemical periodicity that was, overall, satisfactory. For example, the configuration of lithium is 2, 1, while that of sodium, which lies in the same group chemically, is 8, 2, 1. Their chemical similarity can thus be attributed to their analogous outer-shell electron arrangements. In the same way, beryllium and magnesium, which belong together in group II of the periodic table, have two outer-shell electrons in Bohr's scheme. Here then lies the origin of the modern notion that atoms fall into the same group of the periodic table if they possess the same numbers of outer-shell electrons, an idea that had already been hinted at by J. J. Thomson.[8]

Bohr then turned to other questions and did not revisit the issue of chemical periodicity for nine years. It was left to various chemists and physicists to formulate improved electronic version of the periodic table.

BOHR'S SECOND THEORY OF THE PERIODIC SYSTEM

In 1921, Bohr returned to the problem of atomic structure and the periodic table. In 1922 and 1923, he published new, more detailed versions of his electronic configurations.[9] In the meantime Arnold Sommerfeld had generalized Bohr's 1913 theory of the hydrogen atom by introducing the notion of elliptical, rather than merely circular, orbits. In the course of this research, which produced the Bohr-Sommerfeld model of the atom, it became clear that two, rather than just one, quantum numbers would be necessary to specify the degrees of freedom of the electron in the hydrogen atom. Bohr promptly applied this reasoning to multielectron atoms, just as he had extrapolated his 1913 theory to more complicated atoms than the theory had originally been devised to explain. He continued to employ the *aufbauprinzip* to build up successive atoms in the periodic table, but now used two quantum numbers: n, the principal quantum number, and k, the second or azimuthal quantum number (fig. 7.3).

H	1				
He	2				
Li	2	1			
Be	2	2			
B	2	3			
C	2	4			
N	2	4	1		
O	2	4	2		
F	2	4	3		
Ne	2	4	4		
Na	2	4	4	1	
Mg	2	4	4	2	
Al	2	4	4	2	1
Si	2	4	4	4	
P	2	4	4	4	1
S	2	4	4	4	2
Cl	2	4	4	4	3
Ar	2	4	4	4	4

Figure 7.3. Bohr's 1923 electronic configurations based on two quantum numbers. Numbers of electrons in consecutive shells beginning with closest to the nucleus.

According to Sommerfeld's elliptical model of the atom, the angular momentum of an electron would change continually, while the orbit precessed independently of the motion of the electron in its ellipse. The latter precessional motion was stipulated by a second quantum number, which depended on the value of the principal quantum number.

In this revised approach an atom of nitrogen with its seven electrons, for example, would have an electronic configuration of 2, 4, 1. Interestingly, Bohr's more detailed theory of 1922 would seem to have taken a step backward, since contrary to the configuration he had given in 1913, the newer version did not accord well with the experimental fact that nitrogen forms three chemical bonds—a point that would eventually be taken up by the main subject of this chapter, namely Edmund Stoner, and which had also been discussed by Main Smith.

In 1922 Bohr gave a series of seven lectures at the University of Göttingen that became known as the Bohrfest. Heisenberg, Pauli, Sommerfeld, and Born among others attended these lectures, in the course of which their semiempirical nature became increasingly evident and the audience demanded mathematical justifications for what Bohr was doing.

It also became abundantly clear that Bohr's theory of the periodic table rested on a mixture of ad hoc arguments and chemical facts without any derivations from the principles of quantum theory to which Bohr alluded. According to Heisenberg, for example.

It could very distinctly be felt that Bohr had not reached his results through calculations and proofs but through empathy and inspiration and it was now difficult for him to defend them in front of the advanced school of mathematics in Göttingen.[10]

A little later, in a book containing Bohr's famous 1923 paper on the *aufbauprinzip*, Pauli made a revealing marginal remark. In discussing the adding of the 11th electron to the closed shell of 10 electrons, Bohr says, "We must expect the eleventh electron to go into the third orbit."

Meanwhile, Pauli's comment in the margin is, "How do you know this? You only get it from the very spectra you are trying to explain!"

ADIABATIC PRINCIPLE

However, it would be incorrect to say that Bohr had no physical basis whatsoever for his *aufbauprinzip*, and the way in which he applied it to multielectron atoms. For example, he often claimed that it was based on a principle of the early quantum theory called the adiabatic principle, developed by the Dutchman Paul Ehrenfest,[11]

> Suppose that for some class of motions we for the first time, introduce the quanta. In some cases the hypothesis fixes completely which special motions are to be considered as allowed. This occurs if the new class of motions are derived by means of an adiabatic transformation from some class of motions already known.

The adiabatic principle allowed one to find the quantum conditions when an adiabatic or gradual change was imposed on a system. It depended on the possibility of deriving the new motion from one that was known by means of a gradual or adiabatic transformation. For example, if the quantum states of a particular system are known, the new quantum states that result from a gradual application of an electric or a magnetic field, can be calculated. The quantities that preserved their values after such a change were known as adiabatic invariants. Such an adiabatic principle was initially thought to apply to simply periodic systems, meaning those for which the two or more mechanical frequencies were rational fractions of each other. Such systems behave in such a manner that the same motion repeats itself after a certain time. The principle was later shown to be applicable to more general multiply periodic systems, whose frequencies are not fractions of each other and for which the motion does not recur. The hydrogen atom as regarded in the Bohr-Sommerfeld model, in which the electron possesses two degrees of freedom, provides one such example of a multiply periodic system. Each separate motion is periodic but there is no simple rational relationship between the frequencies of the two electron motions.

However, in the case of a multielectron atom, even this situation does not hold since one is dealing with an a-periodic system. In such cases the adiabatic principle does not strictly apply. One cannot therefore use it to fix the quantum states of a multielectron atom in a rigorous fashion. Bohr was well aware of this limitation but hoped that a more general proof might be found to justify using the adiabatic principle in such systems. He explicitly addressed this situation:

> For the purposes of fixing the stationary states we have up to this point only considered simply or multiply periodic systems. However the general solution of the equations frequently yield motions of a more complicated character. In such a case the considerations previously discussed are not consistent with the existence and stability of stationary states whose energy is fixed with the same exactness as in multiply periodic systems. But now in order to give an account of the properties of the elements, we are forced to assume that the atoms, in the absence of external forces at any rate always possess sharp stationary states, although the general solution of the equations of motion for the atoms with several electrons exhibits no simple periodic properties of the type mentioned.[12]

Later in his 1923 article he states:

> We shall try to show that not withstanding the uncertainty, which the preceding conditions contain, it yet seems possible

even for atoms with several electrons to characterize their motion in a rational manner by the introduction of quantum numbers. The demand for the presence of sharp, stable, stationary states can be referred to in the language of quantum theory as a general principle of the existence and permanence of quantum numbers.[13]

Bohr therefore seems to ignore the problems that he himself elaborates, and merely expresses the hope of retaining the quantum numbers even though one is no longer dealing with multiply periodic systems.

The main feature of the building-up procedure, was the assumption that the stationary states would also exist in the next atom in the periodic table, obtained by the addition of an additional electron. Bohr assumed that the number of stationary states would remain unchanged from an atom of one element to the next, apart from any additional states belonging to the newly introduced electron. He thereby envisaged the existence of sharp stationary states, and their retention on adding both an electron and a proton to an atom.

This hypothesis, which became known as the permanence of quantum numbers, came under attack from the analysis of atomic spectral lines under the influence of a magnetic field.[14] An atomic core consisting of the nucleus and inner-shell electrons showed a total of N spectroscopic terms in a magnetic field. If an additional electron having an azimuthal quantum number k were to be added, the new composite system was be expected to show $N(2k-1)$ states, since the additional electron was associated with $2k-1$ states. However, experiments revealed more terms. The observed terms were found to split into one type consisting of $(N+1)(2k-1)$ components, and a second type consisting of $(N-1)(2k-1)$ components, adding to a total of $2N(2k-1)$ components. This represented a violation of the permanence of quantum numbers, in view of the two-fold increase in the number of atomic states on the introduction of an additional electron. Bohr's response was to maintain adherence to the permanence of quantum numbers even in the face of this contrary evidence. He then rather mysteriously alluded to a nonmechanical "constraint," or Zwang as he called it, to save the quantum numbers.[15]

So much for Bohr's own attempts to develop a consistent and accurate account of the electronic configurations of many-electron atoms. There were a number of technical difficulties as we have seen, of which he and others were well aware. It would have to wait for a completely unknown graduate student by the name of Edmund Stoner before matters could advance any further forward.

It is not known whether Edmund Stoner was fully aware of all these further subtleties in Bohr's thinking. What seems to be clear is that Stoner did not allow such deep theoretical questions to interfere with his own attempt to give a more detailed account of the spectra of multielectron atoms as will be discussed in the next section.

STONER'S PAPER

Edmund Stoner's article begins rather boldly:

> The scheme for the distribution of electrons among the completed sub-levels in atoms proposed by Bohr is based on somewhat arbitrary arguments as to symmetry requirements; it is also incomplete in that all sub-levels known to exist are not separately considered. It is here suggested that the number of electrons associated with a sub-level is connected with the inner quantum number characterizing it, such a connection being strongly indicated by the term multiplicity observed for optical spectra. The distribution arrived at in this way necessitates no essential change in the process of atom-building pictured by Bohr; but the final result is somewhat different in that a greater concentration of electrons in outer sub-groups is indicated.[16]

Stoner proceeds to examine the classification of the lines in the X-ray spectra of the elements and points out that Landé has recently provided a successful scheme which draws on the use of the inner or j quantum number in addition to the first two quantum numbers of n and k (denoting the length of the major axis of elliptical paths, and the eccentricity, respectively), in the Bohr-Sommerfeld model of the atom. In addition, as Stoner mentions, this scheme has the virtue of highlighting the analogy between the X-ray and optical spectra obtained from samples of all elements. Nevertheless, Stoner believes that it invalidates the interpretation of certain doublets of spectral lines, the so-called relativity doublets, for which there is ample experimental evidence and a solid theoretical interpretation.

What Stoner was criticizing in his opening remarks was a table of configurations for the noble gas atoms as proposed by Bohr in his 1922 article as shown in figure 7.3. As can be seen Bohr's configurations are completely symmetrical in the sense that equal numbers of electrons are distributed in the same sublevels for any particular n quantum value, or the levels K through P as shown. Furthermore Bohr's scheme commits him to complete rearrangements of sub-level populations that are indicated by the arrows in figure 7.4. For example, the M level consists of two sublevels each containing 4 electrons as in the case of the argon atom. However, in the configuration for the subsequent noble gas atom krypton, that contains 18 outer electrons in the L level, the electrons have rearranged themselves with 6 electrons occupying each of three sublevels. To Stoner both of these features, the symmetry assumed by Bohr as well as the supposed rearrangements of electron groups appeared to be unwarranted by the spectroscopic evidence.

Stoner claimed that the rearrangement of underlying groups occurs in a more natural manner in his own scheme (figs. 7.5 and 7.6). For example, Stoner points out that in his scheme the transition metals

Element	Atomic Number	K	L		M			N				O			P	
							Sub-Levels. $(n, k.)$								
		K	L		M			N				O			P	
		1	2	2	3	3	3	4	4	4	4	5	5	5	6	6
		1	1	2	1	2	3	1	2	3	4	1	2	3	1	2
He......	2	2														
Ne	10	2	4	4												
A	18	2	4	4	(4	4)										
Kr	36	2	4	4	(6	6	6)	(4	4)							
Xe	54	2	4	4	6	6	6	(6	6	6)		(4	4)			
Nt	86	2	4	4	6	6	6	(8	8	8	8)	(6	6	6)	4	4

Figure 7.4. Bohr's scheme based on two quantum numbers and symmetrical electron arrangements.

starting with scandium appear to occupy 10 electrons, following the occupation of a subgroup of 2 electrons, rather than 4, as required by the Bohr scheme. In modern terms we might say that scandium's final three electrons represent the occupation of d-electron orbitals in addition to the two 4s electrons.[17] Similarly, the occupation of electrons into the levels of what we now term the f-block elements takes place in a far more natural manner in Stoner's scheme than it does in Bohr's.

As Stoner summarized,

The present scheme then, accounts well for the chemical properties; it differs from Bohr's in the final distributions suggested, and in the fact that inner sub-groups are completed at an earlier stage, subsequent changes being made by simple addition of electrons to outer sub-levels without reorganization of the group as a whole.[18]

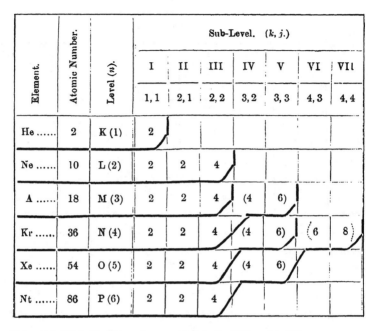

Element.	Atomic Number.	Level (*n*).	Sub-Level. (*k, j*.)						
			I	II	III	IV	V	VI	VII
			1, 1	2, 1	2, 2	3, 2	3, 3	4, 3	4, 4
He	2	K (1)	2						
Ne	10	L (2)	2	2	4				
A	18	M (3)	2	2	4	(4	6)		
Kr	36	N (4)	2	2	4	(4	6)	6	8
Xe	54	O (5)	2	2	4	(4	6)		
Nt	86	P (6)	2	2	4				

Figure 7.5. Table II of Stoner's 1924 article in the *Philosophical Magazine*.

	2 8	8	18	18	36
	$1s^2$ $2s^2 2p^6$	$3s^2 3p^6$	$4s^2 3d^{10} 4p^6$	$5s^2 4d^{10} 5p^6$	$6s^2 4f^{14} 5d^{10} 6p^6$

$_2$He 2

$_{10}$Ne 2. 224

$_{18}$Ar 2. 224. 224

$_{36}$Kr 2. 224. 224. 22446

$_{54}$Xe 2. 224. 224. 22446. 22446

$_{86}$Nt 2. 224. 224. 22446. 22446. 2244668

Figure 7.6. Stoner's scheme in greater detail as compiled by the author. The second line contains the modern configurations for comparison. They draw on 4 rather than 3 quantum numbers.

One of the few pieces of firm evidence that Stoner could point to as favoring his own electron scheme was an experimental study on the spectrum of ionized carbon that had recently been published by R. H. Fowler. Stoner then turned to considering the physical significance of the inner, or j, quantum number that his scheme drew upon. He wrote that, whereas its interpretation had proved very problematical in the context of X-ray spectra, matters became clearer when he considered the analogous optical spectra of the elements. Stoner concluded that the number of possible electrons in an orbit characterized by the quantum numbers n, k, and j should be 2j, an idea that was to prove very useful to Pauli who, as we will see, took up Stoner's article and made it the starting point for what became his own Exclusion Principle.

Stoner turned to considering further experimental evidence in favor of his scheme and pointing against Bohr's. For example, he cited the accurate measurements made on the ratio of two particular X–ray line intensities by Ornstein and Burger. The ratio in question yielded an almost constant value of ½ as one progressed from iron with atomic number 26, all the way to the element tungsten with atomic number of 74. According to Stoner's electron population scheme there were twice as many electrons in the second level as there are in the first, thus accounting perfectly for the relative intensity of ½. On the other hand, Bohr's electron occupation scheme called for there to be three times the number of electrons in the second of these two levels and that would lead to an expected ration of 1 : 3 for the ratio of spectral line intensities contrary to the observations.

MAGNETIC EVIDENCE

Stoner further appealed to measurements on the magnetic properties of metal ions belonging to the first transition series,

The ionic paramagnetism of the third-period elements only will be briefly considered here. The development of ions from K^+ or Ca^{2+} (with 18 electrons) to Cu^+ (with 28 electrons) is brought about, on our view, by the simple addition of electrons to the M_{IV} and M_V sub-levels in the 10 $(4 + 6)$ orbits of the same n,k type $(3,3)$. Sommerfeld, taking into account spatial quantization has shown that the number of Bohr magnetons associated with the ions increases regularly by steps of 1 from 0 to 5 (attaining a maximum value for Mn^{2+} and Fe^{3+} with 23 electrons, and then decreases regularly to 0 (for Cu^+) with increasing numbers of electrons...such a beautiful regularity is in agreement with the development of the M group by the simple addition of 10 similar (n,k) orbits.[19]

To see more clearly what Stoner is claiming, I present in figure 7.7 the electron arrangements of the +2 ions of these metals that have been developed according to Stoner's scheme.

In modern terms the explanation is easily provided by appealing to electrons spinning in opposite directions. From Ca^{2+} up to Mn^{2+}

		Paramagnetism in Bohr magnetons
Ca^{2+}	2,224,2,2,4	0
Sc^{2+}	2,224,2,2,4,1	1
Ti^{2+}	2,224,2,2,4,2	2
V^{2+}	2,224,2,2,4,2,1	3
Cr^{2+}	2,224,2,2,4,2,2	4
Mn^{2+}	2,224,2,2,4,2,2,1	5
Fe^{2+}	2,224,2,2,4,2,2,2	4
Co^{2+}	2,224,2,2,4,2,2,3	3
Ni^{2+}	2,224,2,2,4,2,2,4	2
Cu^{2+}	2,224,2,2,4,2,2,4,1	1
Cu^+	2,224,2,2,4,2,2,4,2	0

Figure 7.7. These electron arrangements are for the 2^+ ions of the transition metals, apart from the final case which is for Cu^+, and coincide with the sequence described by Stoner, from Ca^{2+} up to Cu^+, showing increasing and decreasing values for paramagnetism. (Table prepared by the author).

		Measured values of Bohr magnetons
Ca^{2+}	2, 4,4, 4,4	0
Sc^{2+}	2, 4,4, 4,4, 1	1
Ti^{2+}	2, 4,4, 4,4, 2	2
V^{2+}	2, 4,4, 4,4, 3	3
Cr^{2+}	2, 4,4, 4,4, 4	4
Mn^{2+}	2, 4,4, 4,4, 5	5
Fe^{2+}	2, 4,4, 4,4, 6	4
Co^{2+}	2, 4,4, 4,4, 6, 1	3
Ni^{2+}	2, 4,4, 4,4, 6,2	2
Cu^{2+}	2, 4,4, 4,4, 6,3	1
Cu^{+}	2, 4,4, 4,4, 6,4	0

Figure 7.8. Bohr's configurations for the sequence of ions from Ca^{2+} to Cu^{2+}. Devised by the author based on Bohr's neutral atom configurations.

the number of unpaired electrons increases from zero to five, while from Fe^{2+} up to Cu^+ the subsequent pairing of spins results in there being a decreasing number of unpaired electrons, from 4 in Fe^{2+} all the way to none in Cu^+ (fig. 7.8). Since the very existence of electron spin had not yet been postulated, Stoner's configurations could only truly explain the first half of the sequence, namely from Ca^{2+} to Mn^{2+}. Nevertheless this was a substantial advance on Bohr's scheme that could not even account for the gradual increase in paramagnetism from the Ca^{2+} ion to that of Mn^{2+}.[20]

CHEMICAL PROPERTIES

One of the most convincing set of arguments given by Stoner as to why his scheme was an improvement on Bohr's concerns chemical properties. By consulting figures 7.9 and 7.10 the electron arrangements of the element phosphorus can be compared. Bohr's scheme shows that the five outermost electrons are arranged in two groups consisting of four and one electrons respectively. It follows that Bohr can explain the valency of five shown by phosphorus in some compounds, such as phosphorus penta-chloride, by supposing that all five outer electrons can form five equivalent bonds. Meanwhile Stoner could use his arrangement of outer shell electrons of 2, 2, 1 (fig. 7.10) in order to account equally well for the valency three or five in phosphorus, such as they occur in phosphorus tri-chloride as well as the penta-chloride. As Stoner also pointed out, an analogous situation occurs in antimony the element directly below phosphorus in the periodic table. This element has an electron arrangement of 4, 1 in the case of the Bohr scheme and 2, 2, 1 in Stoner's scheme.

In addition Stoner claimed to explain why the element sulfur can display valencies of 2, 4, or 6 as occur in the di, tetra and hexa-chlorides

H	1						
He	2						
Li	2	1					
Be	2	2					
B	2	2	1				
C	2	2	2				
N	2	2	2	1			
O	2	2	2	2			
F	2	2	2	3			
Ne	2	2	2	4			
Na	2	2	2	4	1		
Mg	2	2	2	4	2		
Al	2	2	2	4	2	1	
Si	2	2	2	4	2	2	
P	2	2	2	4	2	2	1
S	2	2	2	4	2	2	2
Cl	2	2	2	4	2	2	3
Ar	2	2	2	4	2	2	4

Figure 7.9. Based on E. Stoner, The Distribution of Electrons Among Atomic Levels, *Philosophical Magazine*, 48, 719–736, 1924, p. 734.

Stoner's Configurations of 1924 Based on Three Quantum Numbers

Numbers of electrons in successive energy levels beginning closest to the nucleus.						
H	1						
He	2						
Li	2	1					
Be	2	2					
B	2	2	1				
C	2	2	2				
N	2	2	2	1			
O	2	2	2	2			
F	2	2	2	3			
Ne	2	2	2	4			
Na	2	2	2	4	1		
Mg	2	2	2	4	2		
Al	2	2	2	4	2	1	
Si	2	2	2	4	2	2	
P	2	2	2	4	2	2	1
S	2	2	2	4	2	2	2
Cl	2	2	2	4	2	2	3
Ar	2	2	2	4	2	2	4

Based on E. Stoner, The Distribution of Electrons Among Atomic Levels, *Philosophical Magazine*, 48, 719–736, 1924. p. 734

Figure 7.10. Table of configurations based on Stoner's 1924 article, compiled by the present author.

respectively. According to Bohr's arrangement this element could display only two of these three valencies. One can argue that all six electrons would be used in forming bonds or that only the outermost two electrons would be used. It would be far less plausible to argue that the outermost two electrons remained unbounded while the next level of four electrons all entered into bonding in order to account for the formation of the tetra-fluoride and other four-valent compounds of sulfur. As Stoner also mentioned, the same argument holds for compounds of selenium and tellurium that lie below sulfur in the periodic table. In each case his scheme is easily capable of explaining the tetravalent bonding.

Turning to group IV of the periodic table, Stoner could explain the fact that the lower elements, namely tin and lead, are capable of forming di-valent as well as the more common tetra-valent compounds. Finally, in the case of group VII, the variation is even greater in that an element such as iodine shows valencies of +1, +3, +5, and +7, all of which can readily be accommodated by Stoner's scheme except for the mono-valency case. Bohr's scheme as shown in figure 7.3 seems able to explain valencies of +7 and +3 but not +5.[21]

REACTIONS TO STONER'S ARTICLE

In the case of Nicholson (whose work was discussed in an earlier chapter), the reactions were somewhat mixed, ranging from early enthusiastic comments about the accuracy of his calculations to accusations of mere numerology by others such as Rosenfeld. In this respect Stoner's work is rather different in that it was greeted with almost universal acceptance by the experts in atomic physics.

The French physicist De Broglie expressed the view that Stoner's paper was "remarkable," although he identified a few minor quibbles. Sommerfeld, one of the leading players in atomic physics, responded very favorably and included a reference to Stoner in the preface to the fourth edition of his textbook *Atombau*, which he was in the process of completing. It was from *Atombau* that Wolfgang Pauli was to learn of Stoner's work, upon which he would soon capitalize. Sommerfeld wrote that Stoner had cleared up, in a most satisfactory manner, the discrepancy between Bohr's theory and Fowler's experimental observations on the ionized carbon atom. Sommerfeld was quick to extend Stoner's ideas, and in particular the view that the angular momentum of a closed shell or subshell amounted to zero. He further used it to explain why the atoms of magnesium, zinc, and cadmium all displayed a lack of paramagnetism.

Niels Bohr reacted a little more cautiously. While accepting the clear progress that Stoner's scheme represented, he could not abandon what he saw as a need to assign the orbits of electrons unambiguously. Although he did not correspond directly with Stoner, he wrote to Fowler to inform him that he was finding it difficult to take up a definite position on Stoner's work. As late as 1926 Bohr continued to classify his orbits using only the n and k quantum numbers, although in an obvious deference to Stoner he showed the arrangement for neon as $(1_1)^2 \, (2_1)^2 \, (2_2)^6$, by contrast to his earlier more symmetrical arrangement of $(1_1)^2 \, (2_1)^4 \, (2_2)^4$. Coster, while corresponding with Bohr, expressed his support for Stoner although he also mentioned the article's "rather doubtful justification."[22]

A final notable reaction comes from Pauli, who wrote that Stoner's article was "extremely important." A little later he added,

> I am really very enthusiastic about Stoner's paper. The more I read it the more I like it. That was an eminently clever idea to connect the number of electrons in the closed subgroups with the number of Zeeman terms of the alkali spectra.[23]

This is a case that is rather similar to many of those discussed throughout this book: A "bit-part" player, a graduate student in the case of Stoner, publishes a pivotal piece of work that many of the leading experts applaud. Yet the historical record virtually erases the identity of the minor player, in spite of his having provided a crucial stepping stone. Moreover, the fact that such stepping stones are coming from people outside the mainstream of the field highlights the fact that progress is an essentially unified process, with the minor players providing what in evolutionary terms might be called the missing links. More mundanely, the outsiders are able to make these bridging and catalytic leaps because they are relatively unencumbered by what should and should not be done within any particular field. The work of Stoner seems to provide an excellent example of this.

HOW STONER'S ARTICLE LED PAULI TO THE EXCLUSION PRINCIPLE

In his diary Stoner later wrote the following statement about his article on electron arrangements:

> Probably no other single paper of mine has attracted so much attention. This is hardly surprising, for the theme was one of the most basic and topical interest to the chemists as well as physicists, and in essentials it has stood the test of time. Later it would have been presented differently, for at the time of writing the paper neither electron spin nor quantum mechanics had been born. It is of interest to note however, that an explicit statement is effectively made of what later became the Pauli Exclusion principle, though it is presented more as having been arrived at inductively from experimental findings rather than as a basic axiom for a deductive treatment of electron distribution as in Pauli's paper in the following years.[24]

How then did Stoner's article anticipate Pauli's principle?

I do not have the space here to give a detailed account of the factors leading up to Pauli's famous article introducing the Exclusion Principle.[25] Suffice it to say that Pauli had struggled in vain to bring order to the various coupling schemes that had been introduced into atomic physics, with the aim of understanding the details of the atomic spectra as well as the changes that were observed under the influence of magnetic fields. One especially intractable problem had been the so-called anomalous Zeeman effect.[26] After many heroic attempts, Pauli finally informed his colleagues that he was retiring from atomic physics for a while since the field had become too perplexing, even for him. It would appear that Stoner's article was one of the main motivations for Pauli's triumphant return to the problems of atomic physics in 1924. Drawing on several hints, and following a relativistic calculation that received experimental support from Landé,[27] Pauli was able to break with the wisdom of the time in declaring that the valence electrons rather than those in the core were responsible for the main features of the observed atomic spectra. In addition, this view allowed him to free physics from the *Zwang* or nonmechanical constraint that Bohr had been forced to invoke as a stopgap, in order to maintain his building-up principle for atoms in the periodic table. As Heilbron later expressed the matter in his detailed historical account of the Exclusion Principle,

> By piling the sins on his new scapegoat [the valence electron], Pauli permitted bound electrons to retain their identities in the spirit of Bohr's principle of the permanence of quantum numbers.[28]

The transfer of angular momentum from the atomic core to the valence electron that Pauli proposed also led to a more precise assignment of electrons to subgroups than Bohr and even Stoner had been able to achieve. Paradoxically, although Pauli took great encouragement from Stoner's article, he was to drop all mention of the inner, or j, quantum number that had been at the center of Stoner's breakthrough. But while Pauli did away with the j quantum number he was simultaneously elevating Stoner's insight into a principle, the Exclusion Principle, concerning individual electrons in any atom. In the presence of a magnetic field, said Pauli, any valence electron in an alkali atom can take up $\Sigma 2(2k-1)$ or $2n^2$ orbits or the same as the number of orbits in a period characterized by any particular n quantum number. In doing so Pauli added his own new quantum number of $\mu = +1/2$ or $-1/2$, or as it eventually became known, spin of $+1/2$ or $-1/2$. Pauli's "extremely natural rule" as he called it, was simply stated as

It shall be forbidden for more than one electron with the same value of the n quantum number from having the same values for the remaining three quantum numbers.

where Pauli was including his newly introduced fourth or μ quantum number.[29] Pauli's interpretation of his new quantum number μ was that it represented a "non classically describable duplicity" in the quantum mechanical properties of the valence electron.[30]

The fact remains however that Sommerfeld's, and later Stoner's, j quantum number dropped out of the picture. Although Pauli had jettisoned this aspect he retained the all-important factor of 2 that Stoner had stressed earlier in a somewhat different context. This is how I believe we must interpret the opening quotation from this section in which Stoner appears to be taking full credit for Pauli's principle.

Would Pauli or anybody else have arrived at the Exclusion Principle had it not been for Stoner's work? Most probably yes, but the fact remains that Stoner was one of the main catalysts to Pauli's far more celebrated contribution. How much credit should be given to Stoner? These questions and others like it are somewhat superfluous to the present project, which does not seek to apportion credit to one or other person. On the contrary, as I argue throughout this book I merely seek to show the essentially organic nature of scientific progress and the fact that many individuals contribute to varying degrees to the overall evolution of what I call the "living organism" that constitutes science. Stoner's ideas might be regarded as a short-lived intermediate biological species, in much the same way perhaps. Alternatively they could be regarded as an integral part of the overall scientific account of how atomic physics developed, just prior to the discovery of the fully mature quantum mechanics in the years 1925–26.

PRIORITY QUESTION BETWEEN MAIN SMITH AND STONER

I would like to return to the work of the chemist John Main Smith from the previous chapter. As we saw, Main Smith concluded that Bohr's arrangements of electrons were at fault in that they should not be symmetrical. Whereas Bohr had proposed that the shell consisting of 8 electrons should contain a symmetrical arrangement of 4, 4 electrons, Main Smith concluded that it should be a less symmetrical 2, 2, 4. He had arrived at this conclusion from his detailed knowledge of the chemical behavior of the elements, something that Bohr had lacked given his background in physics. In addition Main Smith had drawn on X-ray spectral evidence, which also called Bohr's groupings into question.[31]

What Stoner achieved was essentially a rediscovery of the work of Main Smith, but Stoner received a good deal more credit that Main Smith did. The obvious reason seems to have been that Main Smith chose to publish his work in a chemical journal, and one specializing in industrial chemistry at that. Meanwhile Stoner published his work in a mainstream physics journal and as we have seen was cited by some major players in atomic physics.

Main Smith did not acquiesce entirely to this situation, although he readily acknowledged that Stoner had arrived at his own electron arrangements in an independent fashion. Like Stoner, Main Smith had concluded that the number of electrons in each subgroup was twice the value of a quantum number that was termed the inner quantum number. As a result Main Smith's arrangements were precisely the same as Stoner's. Nor was Main Smith's achievement entirely ignored by the physics community, since Sommerfeld quoted Main Smith's detailed book of 1924 in his own influential article of January 1925.[32]

Main Smith still believed that his contribution had not been sufficiently acknowledged and he wrote to the editor of *Philosophical*

Magazine, stating that several electron distributions that he elabo-rated, "did not originate with Mr. E. Stoner, as within recent months various papers published in your journal have suggested."[33]

His conclusion was that

> I do not suggest that Mr. Stoner's work was not independent, but in view of the foregoing there can be no question of my priority in the matter of this distribution of electrons in atoms. On the continent and in America my priority is acknowledged, but in my own country my work has failed even to be cited in your magazine.[34]

Meanwhile Stoner's work was brought to the attention of Bohr by the spectroscopist Coster. As already mentioned, Bohr was not quick to accept the suggested modification although he eventually altered his tables of electron arrangements in accordance of what must surely called the Main Smith—Stoner scheme.[35]

CONCLUSION

There is no doubt that the work of Stoner aided Wolfgang Pauli on the road to his Exclusion Principle. Until he had read and digested that article Pauli had been floundering in almost complete despera-tion. We are told (among other legends that surround Pauli) that a colleague once found him sitting on a park bench looking rather despondent. Upon asking Pauli what was wrong the friend received the response, "who would not be depressed when thinking about the problems surrounding the anomalous Zeeman effect?" After a hercu-lean effort and aided by Stoner's idea of using the third quantum number in the way that has been discussed in this chapter, Pauli real-ized that yet a fourth quantum number was needed in order to specify a further degree of freedom for each electron in an atom. As a result, the Zeeman effect, in addition to a number of other spectroscopic puzzles, could be resolved. The principle that Pauli announced had extremely far-reaching consequences, not least of which was the solution to a problem that had plagued the Bohr model ever since its inception in 1913: Why do the electrons not fall into the nucleus as predicted by Larmor's formula? Pauli's response was that his exclusion principle forbade any more than two electrons from occupying the same atomic energy level. There was no longer any need for the ad hoc stipulation that electrons simply did not fall into the nucleus and could only percolate as far down as the first energy level.

Could Pauli have arrived at his solution without the work of Stoner? Most probably he could and if he had not found the solution somebody else surely would have done so. But as a matter of historical fact Stoner's work provided the all-important key to unlock the stalemate that Pauli was facing. Stoner's work is as much part of the body-scientific as any other piece of scientific research that was conducted at the time. Scientific research proceeds as an interrelated web or via a process that resembles a giant organism masquerading as a multitude of individuals that frequently appear to be in competition with each other, while all the time furthering the common goal of a deeper understanding of nature.

NOTES

1. Stoner quoted in L.F. Bates, "Edmund Clifton Stoner, 1899–1968," *Biographical Memoirs of Fellows of the Royal Society*, 15, 201–237, 1969, quotation is from p. 212.
2. E.C. Stoner, notes on Bohr's "The Quantum Theory of Atomic Structure," MS333/27.
3. Planck's work revealed the quantization of "action," that is to say, energy divided by frequency. This quantity is now of historical interest only, and it is more common to refer to the quantization of energy, or the action of a particular system multiplied by its frequency.

4. Conversely, electrons could undergo transitions to less stable orbits following the absorption of specific quanta of energy.

5. The quantization of angular momentum assumed by Bohr as well as the notion that electrons in stationary states do not radiate was somewhat ad hoc. It was justified later by Erwin Schrödinger's approach to calculating the energy of the hydrogen atom, in which quantization emerges as a consequence of the application of boundary conditions.

6. N. Bohr, "On the Constitution of Atoms and Molecules," *Philosophical Magazine*, 26, 476–502, 1913.

7. Ibid., p. 497.

8. This statement is a simplification and is only correct for the main-group, or representative, elements in the periodic table. In the case of the transition elements, the members of a group of elements have the same number of electrons in the same penultimate shell. In the rare earths, the elements in the same group have the same number of electrons in a shell located two shells from the outer shell. And there are further deviations given that about 20 elements have anomalous configurations,

9. N. Bohr, "Uber die Anwendung der Quantumtheorie aud den Atombau," *Zeitschrift für Physik*, 13, 17–165, 1923.

10. Heisenberg quoted in H. Kragh, "The Theory of the Periodic System," in A.P. French, P.J. Kennedy, eds., *Niels Bohr, A Centenary Volume*, Harvard University Press, Cambridge, MA, 1985, 50–67, quotation is from p. 61.

11. P. Ehrenfest, "Adiabatic Invariants and the Theory of Quanta," *Philosophical Magazine*, 33, 500–513, 1917, quotation is from p. 501.

12. N. Bohr, "Uber die Anwendung der Quantumtheorie aud den Atombau," *Zeitschrift für Physik*, 13, 17–165, 1923, quotation is from p. 129.

13. Ibid., p. 130.

14. A. Landé, "Über den anomalen Zeemaneffekt," *Zeitschrift für Physik*, 15, 189, 1923.

15. J. Hendry, *The Creation of Quantum Mechanics and the Bohr-Pauli Dialogue*, Riedel, Dordrecht, Holland, 1984.

16. E.C. Stoner, "The Distribution of Electrons among Atomic Levels," *Philosophical Magazine*, 48, 719–736, 1924, quotation is from p. 720.

17. Stoner's paper represents the first case of such an argument that is nowadays taken completely for granted.

18. E.C. Stoner, "The Distribution of Electrons among Atomic Levels," *Philosophical Magazine*, 48, 719–736, 1924, quotation is from p. 724.

19. Ibid., p. 733.

20. The modern account holds that electrons occupy orbitals in such a way that spins are paired. An ion of Co^{2+} therefore has seven electrons distributed in a set of 5 d-orbitals in such a way as to leave three unpaired electrons and therefore a magnetism corresponding to three Bohr magnetons, as observed.

21. Stoner rather conveniently omits to mention the mono-valency of halogen atoms.

22. Coster letter to Bohr, Dec 7, 1924, in J. Rud Nielsen ed., *Collected Papers of Niels Bohr*, vol. 3, Periodic System (1920–1923).

23. Pauli letter to Landé, 24 November, 1924, *Pauli Briefwechsel, Bd.* 1.

24. L.F. Bates, "Edmund Clifton Stoner, 1899–1968," *Biographical Memoirs of Fellows of the Royal Society*, 15, 201–237, 1969, quotation is from p. 214.

25. This task has already been undertaken by J. Heilbron, "The Origins of the Exclusion Principle," *Historical Studies in the Physical Sciences*, 13, 261–310, 1983. Incidentally on reading this article as well as Daniel Server's paper dealing with a slightly earlier period in atomic physics, one cannot fail to be struck by the gradual and the evolutionary nature of the growth of atomic physics.

    Progress seems to involve a patchwork of ideas coming from many different contributors each reacting to the work of his peers. My use of the word "patchwork" should in no way be taken to indicate support for the views of Nancy Cartwright who uses this term but very much in the context of arguments purporting to show the essential disunity of science. N. Cartwright, *The Dappled World: A Study of the* Boundaries of Science, Cambridge University Press, Cambridge, 1999. My own patchwork is very simply undergirded by what I claim to be the essential and complete unity of science seen as an organic whole.

26. In the normal Zeeman effect the application of a magnetic field to an atomic spectrum results in a splitting of such things as an equally spaced triplet, as occurs for example case of hydrogen. In other atoms however the anomalous Zeeman effect results in the splitting of spectral lines into four, six, or even more lines. Such deviations remained mysterious until Wolfgang Pauli introduced the use of a fourth quantum number to characterize the stationary states of electrons in atoms.

27. W. Pauli, "Uber den Einfluss der Geschwindi gkeitsabhangigkeit der elektronmasse auf den Zeemaneffekt," *Zeitschrift fur Physik*, 31, 373–385.

28. J. Heilbron, "The Origins of the Exclusion Principle," *Historical Studies in the Physical Sciences*, 13, 261–310, 1983, quotation is from p. 302.

29. The four quantum numbers discussed by Pauli were, n, k, m_l, and μ. These days the Pauli Principle is stated in terms of the quantum numbers, n, l, m_l, and m_s but the content remains exactly the same.

30. Again in modern parlance this duplicity is assigned to the two possible values of the spin quantum number m_s. Pauli was rather resistant to the interpretation of the fourth quantum number in terms of any mechanical motion or "spin." He relented a good deal later after the contributions of Uhlenbeck and Goudsmit as well as calculations by Thomas. In some respects he was right to resist since it is now generally agreed that spin is something of a misnomer, given that the property in question does not constitute any form of classical mechanical

motion but rather an intrinsic, and it has to be said, still somewhat mysterious form of angular momentum.

31. Even earlier, the French physicists Louis De Broglie and Alexandre Dauvillier had also challenged Bohr's scheme, on the basis of their own X-ray spectral evidence.

32. Arnold Sommerfeld cited Stoner's article in the preface of the fourth edition of his book, *Atomic Structure and Spectral Lines* (*Atombau und Spektrallinien*), Braunschweig, F. Vieweg & Sohn Akt.-Ges., 1924.

33. J. Main Smith, "The Distribution of Electrons in Atoms" [letter dated September 8], *Philosophical Magazine*, 50(6), 878–879, 1925, quotation is from p. 878.

34. Ibid., pp. 878–879.

35. As mentioned earlier this label was indeed used by the chemist Glasstone in a textbook that he authored.

Chapter 8

Charles Janet

Charles Janet, the final scientist we will explore (fig. 8.1), differs somewhat from the rest in that he did not produce a piece of work that catalyzed the discoveries of others. Nevertheless he does resemble the others covered in this book in having been somewhat obscure, perhaps even more so than the others. However, Janet's work has been rediscovered more recently and has spawned much research on presenting the periodic system in an optimal fashion. Janet was also very much like Anton van den Broek in being a complete outsider to the professional scientific community while still managing to make world-class contributions to the literature.

Charles Janet was born in Paris in 1849, twenty years before Mendeleev published his famous periodic table of 1969. Janet's father attended the Ecole Polytechnique and was a leading statesman in France. Charles the son graduated as an engineer and was involved in many business ventures, among them being the technical director of a factory at Saint-Ouen. While still pursuing his very successful business career Janet began to study at the Sorbonne, eventually obtaining a doctorate in natural science. In 1882, at the age of 33, he published his first scientific article, on the geology of the Parisian Basin. A little later he turned to the subject of insects, specializing in bees and ants, a topic on which he continued to publish throughout his life. Between 1911 and 1915 Janet also threw himself into a study of fresh water algae. In 1927, when he was 78, he turned to chemistry and produced the work which has been rediscovered in the 20th and 21st centuries and

Figure 8.1   Charles Janet.

which has led to renewed interest into the theoretical foundations of the periodic system of the elements.

What Janet did was to propose an original representation of the periodic system that possesses great formal beauty and apparent coherence with quantum mechanics, although much controversy has also been raised by this last proposal. Janet produced the so-called left-step table that resembles a staircase rising from left to right, unlike the more uneven format in which the periodic table is normally displayed. According to some authors, Janet's table is more in keeping with quantum mechanics which is generally believed to provide the theoretical foundation for the system of the elements.[1]

Janet's published corpus comprises a staggering 4000 pages and 700 figures. In addition he assembled a fossil collection with over 50,000 items that he personally classified into 400 or so species. Janet also had a number of very different interests which included making electrical inventions, the structure of icebergs, and the housing conditions of early twentieth-century French laborers. As his biographer, Loic Casson, has described him, Janet was a savant in the literal sense of one possessing an encyclopedic knowledge of various quite different fields of study.[2]

Janet's work gained the recognition of many learned societies even within his own lifetime. In 1896 he was awarded the Prix Thore by the French academy of science for his work on the anatomy of red ants. In 1899 he was elected president of the Zoological Society of France. He also became well known in zoological circles for his invention of an artificial nest, which allowed biologists to observe the social behavior of ants. In 1910 Janet was awarded the Cuvier prize for the best zoological work of the year in France. Although Janet gained a good deal of recognition among entomologists, zoologists, and biologists generally, he was forgotten in his hometown of Bauvais. As his biographer notes, there is no school, building or road named after him.

Janet remained much better known as the director of La Brosserie Dupont, one of the largest commercial enterprises in the l'Oise district of France, and one which employed about 1000 workers. Meanwhile, Janet lived in a 54-room mansion which survived until it was demolished in 1972. His scientific library was sold soon after his death, as was his enormous fossil collection. It appears that none of his descendants were interested in continuing any of his academic work.

Although Janet worked in the four main areas of geology, entomology, biology, and chemistry, he did not persist in any of these fields for more than about ten years at a time, perhaps contributing to his lack of recognition during his lifetime and beyond. For example, he was never elected to the prestigious Academie des Science.

JANET ON ATOMIC STRUCTURE AND CHEMISTRY

In his detailed and highly insightful study of Janet's periodic systems, Philip Stewart begins a section on the atomic nucleus with the following statement:[3]

> The first of his papers on the elements (Janet 1927a, b) do not augur well. Basing himself on a naïve concept of the atomic nucleus, he attempts a theoretical model of its structure, imagining it as a disk with concentric rings of particles. Four rings would constitute the 'protero-isotope', the supposed lightest nucleus of a given element, 'which might be real and stable or virtual and unstable'. Inside this, additional proton-electron pairs (referred to henceforth as 'neutron-equivalents') would account for the mass of heavier isotopes. The only authors he cites, without references, are Rutherford and Moseley. (Stewart, 2010)

Even though Stewart is technically correct to point out Janet's "mistakes" it may be well to remember that in the spirit of this book we will refrain from making such judgment calls, since as we have seen in the case of Nicholson in particular, many apparent mistakes can still lead to scientific progress. Alternatively as I have also tried to suggest, there is little room for viewing scientific developments as being right or wrong. They are just developments, sometimes partial, sometimes able to stand the test of time, and all very much like evolutionary developments that take place within living organisms, where one would never dream of labeling any such developments as being either correct or otherwise.

In the case of Janet, the fact remains that from his bizarre-seeming views about the nucleus he quickly arrived at a revolutionary view of the periodic system which continues to challenge chemists and physicists up to the present and which may even provide the optimal representation of the periodic system in due course. Even though Philip Stewart points out what he regards as Janet's early mistakes it is clear that he holds an immense admiration for the French savant, and even goes as far as to suggest that Janet anticipated the existence of deuterons and helium-3 nuclei and also the notion of nucleosynthesis in the interior of stars as later elaborated by Burridge Burridge and Fowler and Hoyle[4]—or as the their famous article is colloquially known, B²HF.

JANET ON THE PERIODIC SYSTEM

What has subsequently become known as Janet's left-step periodic system first appeared in an article published in 1927 (fig. 8.2).[5] However, this first version lacks the main characteristic of Janet's left-step table, as it is discussed in the contemporary literature, namely the placement of helium at the top of the alkaline earth group of elements rather than at the top of the noble gases. (However Janet announced that he would make this controversial move in a future article.) What he does do in the 1927 article is to separate the actinide elements (Ac, Th, Pa, U), from the main body of the table. In doing so Janet achieves a step that is generally attributed to the American chemist Glen Seaborg—who only proposed this change to the periodic system in the mid 1940s. Here is one development that is indisputably anticipated by Janet.

It is also in this article that Janet gives the first hint of what has become his main contribution to science, namely the apparently bizarre notion of moving the element helium from the noble gas group to the head of the alkaline earth elements, as shown in figure 8.3. In Janet's own words,

> Le troisième de ces Essais, que nous exposerons dans un prochain Fasicule, est celui qui, de beaucoup, nous paraît le meilleur. Il serait

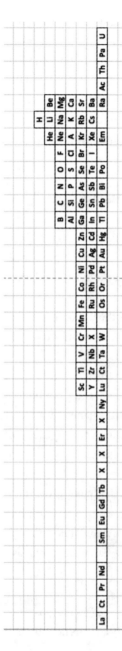

Figure 8.2. The first periodic table to show the elements in a left-step format. The table also clearly shows that Janet realized the need to remove the actinide elements from the main body of the table. Previous authors had considered the elements Ac, Th, Pa, and U as analogues of the transition metals, La, Hf (or Ct as shown here), Ta, and W. Redrawn by author with original element symbols used by Janet retained.

probablement facile de le faire admettre, n'était ce fait qu'il indique que l'Hélium n'appatiendrait pas à la colonne des Gaz rares, comme c'est universellement admis, mais, bien, à la colonne des éléments alcalino-terreux. Ce déplacement sugerré par la parfaite régularité géométrique et arithmétique du Tableau auquel il conduit nous a, tout d'abord, semblé si téméraire, que nous avons hésité à l'admetttre. Cependant, il parait être confirmé, comme nous l'indiquons page 89, par la structure du système planétaire de l'atome de l'Helium, structure qui concorde bien mieux, avec celle des éléments alcalino-terreux, qu'avec celle des Gaz rares.

My translation:

The third of these attempts, which we will display in a future article, is the one that appears to us to be by far the best one. It would probably be easy for it to be adopted were it not for the fact that it suggests that helium would not appear as a noble gas, as is generally believed, but instead in the column for the alkaline-earth metals. This displacement, as suggested by the perfect geometrical and arithmetical regularity of the table, at first seemed to us so tentative that we hesitated to entertain it. However it appears to be confirmed, as we indicate on page 89, by the planetary system of the atom of helium, a structure that agrees far better with those of the alkaline-earths than with those of the rare gases.

In addition to publishing many variations on this left-step table theme, Janet's work was essentially based on what he termed a helicoidal system in which he stressed the continuinty of the elements by designing by means of a coiled line, or lines, such as in the examples shown in figure 8.4.

Furthermore, Janet states that he believes that the criterion of true periodicity is the possibility of winding the line representing a series

																														1	2
																														3	4
																								5	6	7	8	9	10	11	12
																								13	14	15	16	17	18	19	20
														21	22	23	24	25	26	27	28	29	30	31	32	33	34	35	36	37	38
														39	40	41	42	43	44	45	46	47	48	49	50	51	52	53	54	55	56
57	58	59	60	61	62	63	64	65	66	67	68	69	70	71	72	73	74	75	76	77	78	79	80	81	82	83	84	85	86	87	88
89	90	91	92	93	94	95	96	97	98	99	100	101	102	103	104	105	106	107	108	109	110	111	112	113	114	115	116	117	118	119	120

Figure 8.3. The first example of a Janet left-step table in which the element helium is classified among the alkaline earth elements. Janet, 1928. Redrawn by author, with elements indicated by their atomic numbers as in Janet's version).

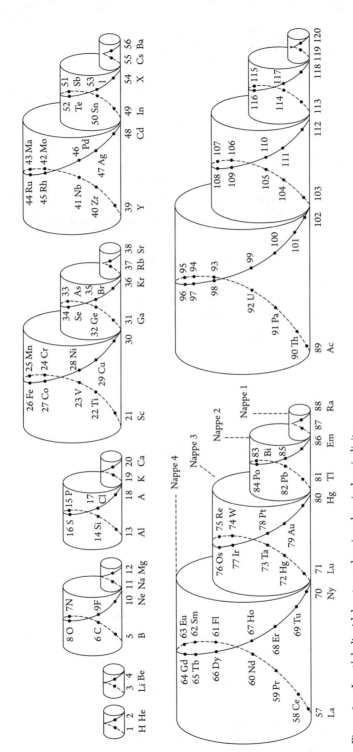

Figure 8.4. Janet's helicoidal systems showing chemical periodicity.

of all the elements into a geometrically defined helix. In this respect he resembles his French compatriot, Emile Béguyer De Chancourtois who is beyond any doubt the first discoverer of the periodic system, although he too received very little credit for his work in his lifetime, or indeed subsequent to his death.[6]

Yet another innovation of Janet's should be mentioned, namely his taking seriously the notion of an element with atomic number of zero and his identification of this element with the neutron. At least one contemporary expert on the periodic table is a supporter of this identification (Stewart, 2010).[7] In a later article of 1930 Janet began to make explicit connections between his own periodic system (arrived at in his idiosyncratic manner) and the account of the periodic table that was being presented by his contemporary atomic physicists including Bohr and Stoner.[8]

THE MADELUNG RULE

It has also been argued that Janet was the first to arrive at what subsequently became known as the Madelung rule, which is intended to denote the order in which electronic orbitals are filled. The rule can be expressed by

$$n + l$$

where n denotes the main quantum number and l the value of the second or angular momentum quantum number. Electrons are frequently said to occupy orbitals according to increasing values of the sum of $n + l$. So for example, the 1s orbital fills before 2s, which in turn fills before 2p and so on. The rule correctly predicts that, in the case of the elements potassium and calcium, the 4s orbital fills before the 3d orbitals since the value of $n + l$ the 4s orbital is 4 as opposed to

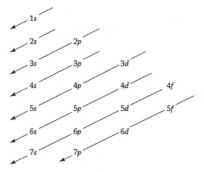

Figure 8.5. Madelung Rule displayed in diagram form. The order of filling is generally said to be provided by following the arrows from right to left, starting at the top with the 1s orbital and proceeding downward. This diagram and many like it are essentially the same as the diagram published by Janet in 1928.

5 for the 3d orbitals (fig. 8.5). Interestingly some recent work has revealed that the Madelung, or perhaps we should be saying the Janet, rule, strictly breaks down as from element 21 (scandium).[9] In this and subsequent transition elements experimental evidence reveals that 3d orbitals are preferentially occupied.

Nevertheless the application of the Madelung/Janet rule still provides the correct overall configuration of atoms of the transition metal series regardless of the precise order of occupation. For example scandium has electronic configuration of $[Ar]\ 3d^1\ 4s^2$ whereas the Madelung-Janet rule gives $[Ar]4s^2\ 3d^1$.[10]

JANET'S LEFT-STEP TABLE IN CONTEMPORARY SCHOLARSHIP ON THE PERIODIC TABLE

Any author who has taken a serious interest in the foundations and significance of the periodic table cannot fail to be struck by the sublime elegance and beauty of Janet's left-step system. Even though he arrived at his system independently of quantum mechanics, Janet's

																														H	He
																														Li	Be
																					B	C	N	O	F	Ne	Na	Mg
																					Al	Si	P	S	Cl	Ar	K	Ca
													Sc	Ti	V	Cr	Mn	Fe	Co	Ni	Cu	Zn	Ga	Ge	As	Se	Br	Kr	Rb	Sr
													Y	Zr	Nb	Mo	Tc	Ru	Rh	Pd	Ag	Cd	In	Sn	Sb	Te	I	Xe	Cs	Ba
La	Ce	Pr	Nd	Pm	Sm	Eu	Gd	Tb	Dy	Ho	Er	Tm	Yb	Lu	Hf	Ta	W	Re	Os	Ir	Pt	Au	Hg	Tl	Pb	Bi	Po	At	Rn	Fr	Ra
Ac	Th	Pa	U	Np	Pu	Am	Cm	Bk	Cf	Es	Fm	Md	No	Lr	Rf	Db	Sg	Bh	Hs	Mt	Ds	Rg	Cn	Nh	Fl	Mc	Lv	Ts	Og	119	120

Figure 8.6. Modern version of Janet's left-step periodic system.

later articles show that he understood the work of Bohr and others in this area.

If one considers an updated version of Janet's left-step table as shown in figure 8.6 one begins to see the beauty of his creation. First the various blocks of the table are displayed in what could be described as a logical order—in order of increasing distance from the nucleus, reading from left to right. For example a 4f orbital electron moves at a smaller average distance from the nucleus than does a 4d electron than does a 4p electron than does a 4s electron.[11] Whether this is a physically significant virtue or not is not so easy to answer however. Since the occupation of atomic orbitals by electrons is a matter of relative energies rather than relative distances from the nucleus there is no absolute necessity for a periodic system to reflect the order of size of orbitals.

However, there is another apparent virtue in the left-step table: As a result of placing helium in the alkaline earth group one is being more faithful to the underlying physical description of atoms, namely in terms of their electronic configurations. An atom of helium has two outer electrons and as a result it would seem to belong quite naturally to the alkaline earth group which also features atoms such as beryllium, magnesium, and calcium, all of which similarly possess two outer shell electrons. The opposing argument from a traditionalist would be that helium can equally well be regarded as a noble gas since its two electrons constitute a full shell and atoms with full outer shells should be regarded as noble gases. It appears that the electronic configuration of helium is ambiguous and does not allow one to categorically settle the question of which group it should be placed in. If we focus on the number of electrons present it should be regarded as an alkaline earth element, whereas if we focus on the number of electrons required to fill its outer shell, helium is then a noble gas.

Turning to yet another apparent virtue in the left-step table that features helium in the alkaline earths, we can focus on the lengths of successive periods. In the left-step table every single period length repeats once to give successive periods of 2, 2, 8, 8, 18, 18, 32, and 32 elements. This is the kind of arithmetical regularity that Janet appeals to in the above-cited passage from his 1928 article. Once again this argument is not categorical since a traditionalist can still resist the move by arguing that nature does not have to be quite so regular and that there is nothing inherently wrong with having one of the periods, namely the very first one, not repeating.

Finally we must turn to chemical properties and ask how Janet's left-step table stands in chemical terms. After all, one might argue, the periodic table is primarily intended as a classification of chemical entities (the chemical elements), irrespective of their ultimate physical constitution. Anybody pursuing such a strictly chemical way of thinking might find the notion of helium as an alkaline earth element to be something rather repugnant. Meanwhile, to keep helium in its traditional position at the head of the noble gases would seem to be the eminently sensible decision in chemical terms. Helium is after all an example of a noble gas in the sense of being highly unreactive. Indeed it is the most unreactive of the noble gases and would seem to epitomize the characteristic inertness of this group. Why then should a chemist want to accept the left-step table?

In fact there is a predominantly chemical response to this question but this requires a digression into the philosophy of chemistry and the very meaning of the concept of elementhood; as given in the following section.

ELEMENTS AS BASIC SUBSTANCES OR AS SIMPLE SUBSTANCES

There exists a long-standing puzzle in chemistry that has not received much attention and remains unknown to most professional chemists and chemical educators. This is the dual sense of the concept of an "element." First there is the sense of "element" to denote a simple substance that cannot be decomposed any further and that can exist in isolation such as yellow sulfur or green gas chlorine or the dull silvery metal sodium. Second, there is a more philosophical, more fundamental use of the term to denote the substance that persists across chemical changes and that exists even after, for example, sodium has combined with chlorine. Neither the silvery material generally associated with sodium, nor the green gas associated with chlorine appear to remain when these two substances combine chemically. And yet both "elements" are still present in the form that one influential author in this field has termed "basic substance," meaning the substance that underlies all manifest properties.[12] Some authors go as far as to claim that the element in this second sense is the bearer of properties but is itself essentially devoid of any manifest properties. Other authors have stressed that an element as a basic substance should be characterized just by its atomic number since this quantity is the only one that persists through chemical combination among elements in the first sense of the combination of simple substances.[13]

In historical terms, the ancient Greek philosophers believed that the elements were the bearers of properties—that they were basic substances that stood beneath manifest properties. Such philosophical views appear to have persisted well into the Middle Ages and the alchemical era and may well have contributed to the overall obfuscation that existed until the chemical revolution. One of Lavosier's triumphs during the chemical revolution was his turning away from the more philosophical sense of element to embrace a more positive sense of an element, as something that could actually be isolated, or an element as a simple substance as we are calling it here.

Interestingly Mendeleev, who came a good deal later than Lavosier, can be seen to mark a return toward embracing both senses of the concept of an element. Indeed Mendeleev went to great lengths to stress that the periodic system was above all a classification of the elements as basic substances rather than merely a classification of elements as simple substances. Perhaps this explains why Mendeleev was equally as interested in the properties of isolated elements as he was in the properties of "elements" when they were present in compound form.

> It is useful in this sense to make a clear distinction between the conception of an element as a separate homogeneous substance, and as a material but invisible part of a compound. Mercury oxide does not contain two simple bodies, a gas and a metal, but two elements, mercury and oxygen, which, when free, are a gas and a metal. Neither mercury as a metal nor oxygen as a gas is contained in mercury oxide; it only contains the substance of the elements, just as steam only contains the substance of ice, but not ice itself, or as corn contains the substance of the seed but not the seed itself.[14]

Much has been written on this topic in the newly emerging field of the philosophy of chemistry and I do not intend to dwell on the subject too much longer.[15] Suffice it to say that if we take Mendeleev at his word it makes the notion of regarding helium as an alkaline earth element a little more palatable. The reason is that in this wider sense of what it is to be an element we are no longer obliged to focus on the chemical behavior of helium as a simple substance. The arguments concerning the highly inert nature of helium that are generally taken as supporting its being a noble gas might begin to lose their force. Having said this, as far as we know helium shows no behavior whatsoever in combination with other elements and so this approach does not seem to offer much hope for the Janet-style placement of helium among the alkaline earths but it does at least open the door a little.

As I concluded in a recent book,[16] I don't believe we are yet in a position to resolve the question of the placement of the element helium in the periodic table and as a result the status of Janet's periodic table must remain somewhat in limbo although it continues to challenge our deepest attempts to understand the periodic system and its underlying meaning.

As Philip Stewart has stressed, the work of Janet suffered severely due both to having been privately published and to its being presented in French. There appear to have been several failed attempts to report his work to an English speaking audience that only rendered matters worse. For example, an article supposedly authored by Janet appeared in *The Chemical News* dated 1929. However even a casual inspection of this article shows that it is rather a distorted attempt to present Janet's work and was clearly written by another author—perhaps even the editor of the journal, William Crookes, who had a deep interest in the periodic table. References are omitted or cited inaccurately, figures are mislabeled or absent altogether. Another attempt made to present Janet's work to the Anglophone world was carried out by the chemist L. M. Simmons, himself the author of several influential articles on the periodic system.[17] This attempt appears to be based on the above-mentioned article in *The Chemical News* that was allegedly authored by Janet, thus leading to the introduction of yet further mistakes.

Janet did not fare much better at the hands of the two classic surveys of the periodic system that were published in the 20th century, those of Van Spronsen and Mazurs. By far the better of these two books is the one by the Dutchman Jan Van Spronsen, which highlights Janet's use of spirals as well as nested spirals in his ingenious attempts to represent the periodic system in a graphical manner. The book by Mazurs has frequently been criticized, and rightly so, because this author insisted on redrawing most of the periodic tables he presented from other authors and, in many cases, bringing them up to date by inserting more recently discovered elements and sometimes rotating tables almost at will. Nevertheless Mazurs does declare that Janet is to be considered as one of the four scientists who did the most to establish "good periodic tables."[18]

Finally the 21st century has witnessed a big revival in interest in the work of Janet with articles and books by Katz, Scerri, Bent, and others.[19] Some of the debates among these authors have become so contentious as to cause the editors of the once venerable *Journal of Chemical Education* to declare a complete moratorium on all papers on the periodic system, a move which is clearly rather excessive given the importance of the periodic system to chemistry instructors at every level.

To end on a personal note, I became an ardent supporter of Janet's system for a period of time but eventually grew increasingly suspicious of the chemical plausibility of placing helium among the alkaline earth elements. Instead I proposed a periodic table that was inspired by Janet's left-step arrangement, but instead of moving helium to the alkaline earths I performed the somewhat opposite maneuver of moving hydrogen from its customary position at the head of group 1 to the head of the halogens or group 17 (fig. 8.7). The advantage of such a move is that the periodic table thereby acquires a new atomic number triad in the shape of H (1), F (9), and Cl (17), whereby the atomic number of the middle element has an atomic number that is precisely the mean of the two flanking elements.[20]

The overall result of this change is that the first two elements in the periodic system, hydrogen and helium, both belong to perfect atomic number triads rather than the situation in the conventional periodic table in which only helium does. I also pointed out that any movement of helium out of the noble gases and into another group results in the loss of a perfectly good atomic number triad. Given that the recognition of triads of elements, albeit it atomic weight triads, provided the very first hint of the existence of an ordered system of elements, I argued that having two atomic number triads rather than just one involving the first two elements should be regarded as a strongly supporting factor for this modified periodic table.

Not long after I made this proposal Philip Stewart pointed out that Janet had not only considered such a system but had published it

																												H	He	Li	Be
																								B	C	N	O	F	Ne	Na	Mg
																								Al	Si	P	S	Cl	Ar	K	Ca
														Sc	Ti	V	Cr	Mn	Fe	Co	Ni	Cu	Zn	Ga	Ge	As	Se	Br	Kr	Rb	Sr
														Y	Zr	Nb	Mo	Tc	Ru	Rh	Pd	Ag	Cd	In	Sn	Sb	Te	I	Xe	Cs	Ba
La	Ce	Pr	Nd	Pm	Sm	Eu	Gd	Tb	Dy	Ho	Er	Tm	Yb	Lu	Hf	Ta	W	Re	Os	Ir	Pt	Au	Hg	Tl	Pb	Bi	Po	At	Rn	Fr	Ra
Ac	Th	Pa	U	Np	Pu	Am	Cm	Bk	Cf	Es	Fm	Md	No	Lr	Rf	Db	Sg	Bh	Hs	Mt	Ds	Rg	Cn	Nh	Fl	Mc	Lv	Ts	Og	119	120

Figure 8.7. Variation of left-step table first published by Janet and independently rediscovered by the author (2011), whose motivation was the formation of a new perfect atomic number triad consisting of H, F, and Cl.

in the very same article as his better known "helium moves to the alkaline-earth elements" (Stewart, 2010).

CONCLUSION

The case of Janet is somewhat different from many of the other six scientists featured in this book. But he does share the characteristic of being almost completely unknown while at the same time having catalyzed some very profound developments and debates among many better-known participants.

The evolution of science takes many complicated forms in which being right or wrong matters very little in the overall growth of knowledge. Sometimes the most bizarre thinking succeeds in raising some of the most interesting ideas which can flourish into highly successful scientific discoveries. The case of Janet remains somewhat suggestive but unresolved. This feature does not diminish his contribution which is another example of one made by a complete outsider to the field—not unlike the case of the economist Van den Broek, as discussed in chapter 3.

NOTES

1. Perhaps the leading proponent of Janet's left-step table in recent times has been the recently deceased and well-known inorganic chemist Henry Bent. H.E. Bent, *New Ideas in Chemistry from Fresh Energy for the Periodic Law*, AuthorHouse, Bloomington, IN, 2005. Other sources who have reintroduced Janet's table into the contemporary literature have included, G. Katz, "The Periodic Table: An Eight Period Table for the 21st Century," *The Chemical Educator*, 2001, 6, 324–332; and E.R. Scerri, "Presenting the Left-Step Table," *Education in Chemistry*, September, 135–136, 2005. The classic sources on the periodic table such as van Spronsen and Mazurs also featured Janet's table, as did I in my 2007 book, *The Periodic Table, Its Story and Its Significance*, Oxford University Press, New York, 2007.

2. L. Casson, "Notice biographique sur la vie et de l'oeuvre de Charles Janet," *Bulletin de la Sociéteé. Académique de l'Oise*, 232, Beauvais, 2008.

3. P. Stewart, "Charles Janet: Unrecognized Genius of the Periodic System," *Foundations of Chemistry*, 12, 5–15, 2010, quoted from p. 6.

4. E.M. Burbidge, G.R. Burbidge, W.A. Fowler, and F. Hoyle, "Synthesis of the Elements in Stars," *Reviews of Modern Physics*, 29(4), 547, 1957. Quoted from p.6. A general account of the B²FH article in the context of theories of nucleo-synthesis is given in E.R. Scerri, *The Periodic Table, Its Story and Its Significance*, Oxford University Press, New York, 2007, chapter 10.

5. C. Janet, "La structure du noyau de l'atome considérée dans la classification périodique des Eléments Chimiques," Imprimerie Départementale de l'Oise, Beauvais, 1927.

6. For an account of De Chancourtois' unique contribution to the discovery of the periodic system the reader is referred to E.R. Scerri, *The Periodic Table, Its Story and Its Significance*, Oxford University Press, New York, 2007, chapter 3.

7. This author has gone as far as to propose that Janet anticipated the existence of anti-matter, well before the work of Dirac.

8. C. Janet, "Concordance de l'arrangement quantique de base des électrons plané-taires des atomes avec la classification scalariforme hélicoïdale des elements chimiques." Beauvais Imprimerie Départementale de l'Oise, Beauvais, 1930.

9. W.H.E. Schwarz, "The Full Story of the Electron Configurations of the Transition Elements," *Journal of Chemical Education*, 87(4), 444–448, 2010; E.R. Scerri, "The Trouble with the Aufbau Principle," *Education in Chemistry*, November, 24–26, 2013.

10. The correct order of filling matters when we come to consider the order of ion-ization or de-occupation of the orbitals in any particular atom. If one adheres to the correct order of occupation in scandium, then it's easily understood why the configuration of the singly charged cation of scandium should be [Ar] 3d¹ 4s¹ as experimentally observed. The usual presentations given in chemistry and physics textbooks imply that the Madelung/Janet rule is strictly obeyed in atoms such as scandium with the result that it becomes rather mysterious why ionization should result in the removal of a 4s electron. This has not prevented many textbooks from inventing all manner of ad hoc explanations in which they attempt to have it both ways, namely 4s is preferentially occupied and also preferentially ionized. This is quite simply incorrect as pointed out in a number of articles, such as W.H.E. Schwarz, "The Full Story of the Electron Configurations of the Transition Elements," *Journal of Chemical Education*, 87(4), 444–448, 2010; E.R. Scerri, "The Trouble with the Aufbau Principle," *Education in Chemistry*, November, 24–26, 2013.

11. Janet's system contains a total of 120 elements, which is coincidentally just two elements short of the current heaviest element that has been synthesized, namely element 118. Because of his strong commitment to geometrical and

arithmetic regularity Janet predicts that there will be no elements heavier than element 120, something which is doubted these days. The natural limit to the periodic table is considered to be either 137 or 172–173 depending on the particular approach adopted in making this prediction. See E.R. Scerri, "Cracks in the Periodic Table," *Scientific American*, June, 68–73, 2013.

12. F. Paneth, "The Epistemological Status of the Chemical Concept of Element," *British Journal for the Philosophy of Science*, 13, pp. 1–14 (Part I) and 144–160 (Part II), 1962. Reprinted in *Foundations of Chemistry*, 5, 113–145, 2003.

13. E.R. Scerri, "What Is an Element? What Is the Periodic Table? And What Does Quantum Mechanics Contribute to the Question?," *Foundations of Chemistry*, 14, 69–81, 2012; "The Dual Sense of the Term "Element," Attempts to Derive the Madelung Rule and the Optimal Form of the Periodic Table, If Any," *International Journal of Quantum Chemistry*, 109, 959–971, 2009; "Some Aspects of the Metaphysics of Chemistry and the Nature of the Elements," *HYLE— International Journal for Philosophy of Chemistry*, 11, 127–145, 2005.

14. D.I. Mendeleev, D.I., *The Principles of Chemistry*, Longmans, Green and Co., London [first English translation from the fifth Russian edition], 1981, p.23.

15. In addition to the articles already cited above, other contemporary articles include, J. Earley, "Why There Is No Salt in the Sea," *Foundations of Chemistry*, 7, 85–102, 2005; R.F. Hendry, "Lavoisier and Mendeleev on the Elements," *Foundations of Chemistry* 7, 31–48, 2005; K. Ruthenberg, "Kant, and the Philosophy of Chemistry," *Foundations of Chemistry* 11, 79–91, 2009; Scerri, E.R., "Some Aspects of the Metaphysics of Chemistry and the Nature of the Elements," *HYLE—International Journal for Philosophy of Chemistry*, 11, 127–145, 2005; Scerri, E.R., "The Dual Sense of the Term 'Element,' Attempts to Derive the Madelung Rule, and the Optimal Form of the Periodic Table, If Any," *International Journal of Quantum Chemistry*, 109, 959–971, 2009; "What Is an element? What Is the Periodic Table? And What Does Quantum Mechanics Contribute to the Question?," *Foundations of Chemistry*, 14, 69–81, 2012.

16. E.R. Scerri, *A Very Short Introduction to the Periodic Table*, Oxford University Press, Oxford, 2011.

17. L.M. Simmons, "The Display of Electronic Configuration by a Periodic Table," *Journal of Chemical Education*, 25(12), 658–661, 1948.

18. E. Mazurs, *Graphic Representations of the Periodic System during One Hundred Years*. University of Alabama Press, Tuscaloosa, 1974, p. 141.

19. G. Katz, "The Periodic Table: An Eight Period Table for the 21st Century," *The Chemical Educator*, 2001, 6, 324–332; E.R. Scerri, "Presenting the Left-Step Table," *Education in Chemistry*, September, 135–136, 2005; H.E. Bent, *New Ideas in Chemistry from Fresh Energy for the Periodic Law*, Author House, 2005.

20. E.R. Scerri, "The Past and Future of the Periodic Table," *American Scientist*, 96, 52–58, January-February, 2008.

Chapter 9

Bringing Things Together

In the previous chapters I highlighted the work of seven relatively unknown scientists and suggested that the growth of science takes place in a more organic and interconnected way than is generally believed. I chose to focus on these individuals since their work served to catalyze the work of others. I believe that the seven scientists in question were part of a tacit network of researchers, even though none of them was in direct contact with the leading protagonists of the day—such as Bohr or Pauli. Bohr had briefly met Nicholson but did not think highly of him and did not conduct any correspondence with him. Nor, as far as is known, did Moseley communicate with van den Broek, whose seminal idea concerning atomic number he set out to investigate. Bohr had very little to do with Charles Bury whose work on electronic configurations he very briefly acknowledged. Pauli did not discuss any ideas with either Stoner or Main Smith although he helped himself to some of their key ideas.

The main protagonists made full use of the published work of the seven featured scientists and in some cases acknowledged their contributions. I believe that this kind of activity cuts across the more traditional thinking of science as carried out primarily by outstandingly talented individuals. It speaks to a more entangled, more organic development in which ideas constantly compete and collide with each other and are modified in a trial-and-error fashion rather than through a cold and rational approach. Although I present a less flattering account of the work of the main protagonists I believe it may be a more accurate depiction of what often takes place. My account is more organic and less isolationist, more guided by blind chance and evolutionary forces than by human rationality. Seen from a far distance we might even suppose that science is developing as one large interconnected organism.

Before attempting to gather this material together into a more coherent thesis I propose to explore some other areas in which the traditional image of science, that it is believed to be carried out by highly gifted lone individuals, is put to the test. The following sections of this chapter will examine the occurrence of priority disputes and multiple discoveries. I believe that the notion of a more organic development of science goes a long way toward explaining these two features of science that otherwise remain as somewhat mysterious.

I will then proceed to consider some recent work in the history and philosophy of science, work which seeks to understand how "wrong theories" have sometimes led to some remarkably good science. Again, my organic approach goes some way toward explaining what otherwise seems to be an entirely puzzling aspect of the nature of science.

The remainder of the chapter will focus more directly on an evolutionary growth of scientific theories that I support. I will examine what other authors, including Popper and Campbell, have written on the subject of evolutionary epistemology. I will also consider the writings of Thomas Kuhn on science as an evolutionary process. Contrary to the view of one prominent Kuhn scholar, I argue that Kuhn cannot maintain his early views on scientific revolutions as well as his views on the evolution of science—which were merely hinted at in the final pages of his *Structure of Scientific Revolutions*.

I will review recent work carried out by Kuhn scholars, some of whom conclude that Kuhn's views on revolutions in science changed very considerably. According to these authors, scientific revolutions, for the later Kuhn, were concerned with changes in scientific lexicon and not with the abrupt changes described by the younger Kuhn.

I conclude that my proposed view is not altogether different from that of Kuhn's, minus abrupt revolutions, although I maintain that I arrive at my own view from a somewhat different perspective.

THE NATURE OF SCIENCE AND PRIORITY DISPUTES

Science textbooks typically present theories and concepts as being fully formed, while real science is in a constant state of flux. When science is reported in the press, the errors that led up to a discovery are usually not mentioned. In fact actual science is full of mistakes and wrong turns. Current science is necessarily incomplete and incorrect. One cannot begin to understand the nature of science without facing up to the historical twists, turns, and mistakes that occur. Indeed, the practice of science often involves titanic struggles between individuals or teams of scientists trying to establish their priority, not necessarily because scientists are egotists, but because scientific society rewards the winners and those who can boldly assert their claims. For example, in the search to discover the elements, priority disputes have frequently occurred and in some cases they even continue to occur.[1] One of the most bitter priority issues involved the discovery of element 72 (which was eventually named hafnium).[2]

Arguments and protracted debates, with or without nationalistic undertones, are part of science, whether or not we may like it. Scientific knowledge as a whole might be said to benefit from the fierce scrutiny and competition to which new claims are subjected, even if the individuals involved in the process may suffer as a consequence. But scientific knowledge, in a state of development, is not in the slightest bit interested in the feelings or egos of individual scientists. What matters is the overall progress in human knowledge and not whether the rewards go to one or the other person or nation. Having said that, scientists are humans and scientific knowledge is influenced by various other emotional aspects in addition to nationalism.

Questions of priority have been rampant throughout the history of science. Consider, for example, Darwin versus Wallace, Newton versus Leibniz, Montagnier versus Gallo, or Venter versus Collins. In chemistry, there have been cases like Lothar Meyer versus Mendeleev, Ingold versus Robinson, and H. C. Brown versus Winstein.

An interesting aspect of priority disputes is that people not directly involved in the research frequently seem to take up the cause of a particular scientist and pursue it to a greater degree than the scientists who are directly involved. This was the case in the discovery of several elements, including hafnium, technetium, rhenium, and promethium.[3] In the case of hafnium it was the scientific and popular presses who seem to have made the most vociferous statements on behalf of one or the other of the parties who claimed to have discovered hafnium. To this day there are several chemists and physicists claiming that some of the elements mentioned above were discovered in the early twentieth century and that certain scientists failed to gain credit for one reason or another. Of course, science is full of simultaneous discoveries—as in the discovery of the periodic table itself, usually attributed to Mendeleev alone.[4] Similarly, we have to be prepared for the possibility that many elements were also simultaneously discovered. As Joel Levy points out in his book about scientific feuds,[5]

> The history of science is boring; the traditional version, that is, with its stately progression of breakthroughs and discoveries, inspirational geniuses and long march out of the darkness of ignorance into the light of knowledge. This is the story as it is often presented in museums, textbooks and classrooms; but it is an invention . . .

PRIORITY DISPUTES ACCORDING
TO ROBERT MERTON

Surprisingly little seems to have been written with the aim of analyzing the nature of priority disputes in modern science. One important exception is the work of Robert Merton, the best-known sociologist of science in the classical tradition. In an article written in 1957, he says,[6]

> We begin by noting the great frequency with which the history of science is punctuated by disputes, often by sordid disputes, over priority of discovery. During the last three centuries in which modern science developed, numerous scientists, both great and small, have engaged in such acrimonious controversy.

Merton emphasizes that far from being a rare exception in science, priority disputes have long been "frequent, harsh, and ugly" and that they have practically become an integral part of the social relations between scientists. It would seem to be a simple matter for scientists to concede that simultaneous discoveries often occur and that questions of priority are therefore beside the point. Very occasionally this is just what has happened, as in the cases of Darwin and Wallace, who tried to outdo one another in giving each other credit for their discoveries. A full fifty years after the event, Wallace was still insisting on the contrast between his own hurried work, written within a week after the idea came to him, and Darwin's work, based on twenty years of collecting evidence. While Wallace claimed that he had been a "young man in a hurry," he liked to point out that Darwin had been, "a painstaking and patient student seeking ever the full demonstration of the truth he had discovered, rather than to achieve immediate personal fame."[7]

Merton recounts how in some cases, self-denial goes even further. For example, Euler withheld his long-sought solution to the calculus of variations, until the twenty-three-year-old Lagrange, who had developed a new method needed to reach the solution, could put it into print, so as not to deprive Lagrange. Nevertheless, Merton writes, the recurrent and intense struggles for priority far outnumber these rare cases of noblesse oblige.

Merton presents four possible explanations for the almost ubiquitous current state of priority disputes. First, priority disputes may merely be the expressions of our competitive human nature. If egotism is natural to the human species, claims Merton, then scientists will have their due share of egotism and will sometimes express it through exaggerated priority claims.

His second candidate explanation is that, like other professions, science attracts some, and perhaps many, egocentric people who are hungry for fame. Merton is quickly to dismiss this possibility because he doubts that contentious personalities are especially attracted to science. Rather significantly, Merton recognizes that priority disputes often involve men of modest dispositions who act in assertive ways only when it comes to defending their rights to intellectual property. He also points out that the discoverers or inventors themselves often take no part in arguing their claims for priority—and perhaps even withdraw from such controversies. Instead, it tends to be their followers who take up the cause and regard the assignment of priority as a moral issue that must be bitterly fought over.

Merton believes that by identifying themselves with the scientist or with the nation of which they are a part, these supporters feel that they somehow share in the glory that is obtained if the priority battle is won. After dismissing human nature and the role of supporters in explaining priority conflicts, Merton considers the question of institutional norms,[8]

As I shall suggest, it is these norms that exert pressure upon scientists to assert their claims, and this goes far toward explaining the seeming paradox that even those meek and unaggressive men, ordinarily slow to press their own claims in other spheres of life, will often do so in their scientific work.

Merton finally turns to make what I believe is his most astute point, when he begins to discuss scientific knowledge as a form of property. Whereas the protagonists in commercial disputes can often resolve their differences because there is money to be made, the intellectual property of academics is seldom commercially exploitable.[9] As a result, the only way that the scientist can benefit from his or her "property" is through the fame gained by having made a discovery. Therefore, it is not surprising that scientists will fight ferociously to retain the only benefit that might come from their hard-won intellectual property.

As Merton puts it,

> Once he has made his contribution, the scientist no longer has exclusive rights of access to it. It becomes part of the public domain of science. Nor has he the right of regulating its use by others by withholding it unless it is acknowledged as his. In short, property rights in science become whittled down to just this one: the recognition by others of the scientist's distinctive part in having brought the result into being.

Needless to say, Merton recognizes the large element of nationalism in priority disputes when he writes, "From at least the seventeenth century, Britons, Frenchmen, Germans, Dutchmen, and Italians have urged their country's claims to priority; a little later, Americans entered the lists to make it clear that they had primacy."[10]

A less contentious form of behavior, which is rather more common, is for scientists to fail to cite their competitors. This was true of Mendeleev, the leading discoverer of the periodic table, as many authors have pointed out. Whereas Mendeleev was content to cite the articles of early researchers such as Döbereiner and Pettenkofer, he seems to have been more reluctant to acknowledge the work of immediate competitors such as Lothar Meyer and Newlands, whose work he criticized rather severely. In fact, Mendeleev conducted a rather acrimonious priority dispute with Meyer.[11]

To end this section let me mention some more recent work conducted in the sociology of science and "science studies" following the work of Merton. One interesting example is from Alan Gross (writing in 1998) who says, "I will show that the normative requirements for calling an event a scientific discovery are such that priority and the conflicts it generates are not merely in science, but of science."[12]

Here Gross is again highlighting how scientists often try to deny or at least downplay priority issues as being something that intrudes into the scientific landscape rather than being an integral part of the nature of science. Gross also claims that discovery is not a historical event but rather more in the form of a retrospective social judgment. Finally, he reminds us that Thomas Kuhn had already pointed out that scientific discovery cannot be regarded as a historical event like a war, a revolution, or the crash of the stock exchange.

SIMULTANEOUS OR MULTIPLE DISCOVERY

The history of science also provides us with many instances in which the same discovery was made independently by two or more people at roughly the same time. When we say "independently" it really means that they did not crib from each other or look over each other's shoulder. However, these discoveries cannot be regarded as independent if the state of scientific knowledge as a whole is considered. Science does not just proceed by a series of lucky accidents. It grows in an organic manner, as I have argued, and this inevitably produces many discoveries at each developmental stage.[13]

The notion of multiple discovery in science has a long history but is not generally taken very seriously in scholarly circles. Multiple discovery tends to be mentioned but quickly explained away. This is not surprising in the predominantly individualistic climate that exists in scientific research, despite the rising importance of collaborative research and Big Science.

In fact the denial of multiple discovery is another cause of the all-too-frequent priority disputes that occur in science. As Lamb and Easton stated in the introduction of what is probably the only book-length treatment of the subject of multiple discovery in science,

> However, commentators tend to treat priority disputes as trivial and peripheral to the activity of scientific knowledge. The scientist is therefore in an ambiguous position. On the other hand he is expected to be dedicated to the pursuit of scientific knowledge rather than to the advancement of egotistical claims. Humility is seen as a mark of the scholar. On the other hand the scientific community places a premium on originality, and rewards are dispensed on this basis. Inevitably this results in considerable strain being placed on scientists who are torn between egotism and altruistic impulses.[14]

Whereas the image of scientists is one of the disinterested pursuit of scientific knowledge for the sake of humanity, the reality is frequently one of harsh disputes, accusations of plagiarism, and all-out wars between leading scientists. All of this provides further support for the view that I am proposing here, one of an evolutionary struggle between competing members of a species.

Scientists do not generally like to admit to the occurrence of multiple discovery for a variety of reasons. For example, any acceptance of the possible duplication of research and duplication of knowledge is regarded as being wasteful of increasingly scarce research funding.

Moreover, an admission of multiple discovery or parallel lines of research along very similar lines might raise concerns about possible plagiarism which is of course the worst possible sin that can be leveled at any scientist.

A previous study of multiple discovery was published in 1922 by Ogburn and Thomas, who succeeded in assembling as many as 148 well documented cases.[15] It took a further 50 years before Merton revisited the topic in his book, *Sociology of Science,* and identified 264 cases of multiple discovery.[16] Indeed the gap between these in-depth studies of multiple discovery has been so lengthy that one may almost say that multiple discovery itself has been rediscovered on a number of occasions, albeit separated by long periods of time. According to Merton, there have been as many as 18 such "discoveries" of multiple discovery—to which we should add the work of Lamb and Easton and a number of more recent articles.[17]

TRADITIONAL PHILOSOPHY OF SCIENCE AND MULTIPLE DISCOVERY

Philosophers of science have tended to avoid discussing the phenomenon of multiple discovery. This began with the Logical Positivists who declared that the process of discovery was the concern of psychology, and that their efforts would be directed at examining the logical status of scientific theories rather than their genesis. Karl Popper, for all his well-known opposition to Logical Positivism, seems to have shared this belief in asserting that only the confirmation of theories should be of interest to the philosopher; the initial emergence of theories was to be relegated to mere psychologism and, as such, should be left well alone. How a particular scientist arrived at a particular theory was never something that Popper or his followers were interested in. As Lamb and Easton, along with previous commentators, note, Popper's famous book *The Logic of Scientific Discovery* seems strangely mistitled, given that Popper claims that there is in fact no logic to discovery.[18]

In contrast Lamb and Easton have proposed that the process of discovery is independent of the enquiring mind of any particular individual because it is the result of a collective process, a view with which I have much sympathy. Lamb and Easton call their view one of evolutionary realism and they refuse to draw a distinction between the notions of discovery and invention, believing this to be an outmoded dichotomy. These authors regard both processes as resulting from a collective evolutionary process, very much in keeping with the organic view that I argue for in the present book. To quote Lamb and Easton,[19]

> The search for absolute originality and the quest for priority is misguided, insofar as discoveries can be seen as necessarily multiple or 'in the air'.

They also write,[20]

> We differ from Kuhn insofar as we are committed to an evolutionary and cumulative account of science, rather than a revolutionary model.

However, Lamb and Easton consider that there is a logic of scientific discovery, a claim that I disagree with. They also claim that it would be more appropriate to think of discovery as being explicable rather than being the result of mysterious aspects of individual psychology. Whereas it may be explicable on evolutionary grounds this need not, in my view, imply that it has a logical aspect as such.

But in another remarkable anticipation of my own view, Lamb and Easton write,[21]

In making use of an historicist approach, we advance an evolutionary theory of discovery which likens the growth of knowledge to the development of organic life.[22]

Whereas Popper showed no interest in the process of discovery, Thomas Kuhn, his major rival, believed that discovery has an internal structure. After analyzing a number of cases including the discoveries of oxygen, X-rays, and the planet Uranus, Kuhn also claimed that it is not possible to pinpoint the time and place for any particular discovery. Although I would agree wholeheartedly with such a view it strikes me as odd coming from Kuhn, a well-known advocate of sharp and swift scientific revolutions. One would think that in order to pinpoint a sharp revolution one would equally well need to pinpoint a sharp and clear-cut discovery of one form or another.

POSSIBLE REASONS FOR MULTIPLE DISCOVERY

Several authors have proposed views for the cause of multiple discovery which can be labeled as metaphysical. These authors include the psychologist Karl Jung who appeals to his famous "collective unconscious" to explain the synchronicity of common relationship of all individuals. Likewise, and as reported by Koestler, Buddhist and Taoism philosophical schools seek to unify random events into a grand scheme, adhering to a belief that all things are fundamentally unified in a manner that transcends or undercuts tradition views about causation.

The science scholar De Solla Price takes an altogether different approach in proposing a statistical account of simultaneous discovery. For De Solla Price statistical factors, and the existence of a store of common knowledge among scientists of any particular epoch, together ensure that multiple discovery should be a commonplace phenomenon.

Meanwhile, Merton has this to say:[23]

> Far from being odd or curious or remarkable, the pattern of inde-
> pendent multiple discoveries in science is in principle the dominant
> pattern rather than the subsidiary one. It is the singletons—discov-
> eries made only once in the history of science—that are the residual
> cases, requiring special explanation. Put even more sharply, the
> hypothesis states that all scientific discoveries are in principle mul-
> tiples, including those that on the surface appear to be singletons.[24]

Returning to Lamb and Easton:

> Scientists are bound to the past by their dependence on a deposit
> of accumulated knowledge and bound to the present by their
> interactions with those who share their motivations, cultural
> needs and wants as other members of the scientific community—
> even if they are not always in direct communication with each
> other. Unfortunately multiples are experienced as occupational
> hazards, occasions for stress, disappointment and conflict.[25]

Another author, Benjamin Park goes even further by similarly asserting, "Not an electrical invention of major importance has ever been
made but that the honor of its origins has been claimed by more than
one person."[26]

Lamb and Easton also say,

> The analogy between biological evolution and the evolution of
> invention can be upheld. Neither are absolutely predictable for
> similar reasons. An individual is largely determined by the evolution
> of its species, but from this we cannot predict, with any degree of
> accuracy, the future patterns of evolution....A fundamental char-
> acteristic of organic development is that the system in question

can reach its desired state by a variety of means. Similarly, with inventions, predictions are difficult to formulate because we cannot ascertain precisely the cultural factors involved at any stage of scientific development.[27]

According to Lamb and Easton the process of scientific discovery has a degree of independence from any particular mind as already noted. Accordingly they believe that what matters from the standpoint of evolution is not the success or failure of any particular scientist, but only the overall pattern which emerges from any line of research.[28]

> Evolutionary realism is simply a new way of describing an old doctrine, according to which the analogy between scientific development and organic evolution is linked to the historicist thesis concerning the inter-dependence and relative autonomy of historical structures and particular agents. On these terms scientific development exhibits lines of continuity as well as sharp discontinuities. Scientific theories, world-shattering discoveries and inventions all develop on soil prepared by countless predecessors.[29]

I have to say that I am relieved to find that there is at least one important respect in which my own position differs from that of Lamb and Easton, namely in that, unlike these authors, I reject any significant discontinuities in the history of scientific development.[30] I do not believe that sudden discontinuities occur in the course of biological evolution, except for some rather "artificial" cases such as the extinction of the dinosaurs that had a rather particular cause, namely the huge asteroid which is now believed to have collided with the earth.[31]

THE THESIS OF UNIVERSAL MULTIPLICITY

Some authors who support the view that multiple discovery is a genuine phenomenon have taken the notion to its logical conclusion, that all discoveries are in fact multiples. But the notion of universal multiplicity, as it has been termed, faces some tough challenges such as the obvious objection that it may be unfalsifiable. If one encounters a case that seems to point against multiple discovery a supporter of universal multiplicity would seem to have the option of claiming that if one looks far enough even what appears to be a single discovery will eventually be found to have been multiplied. Lamb and Easton's response to this problem is not to abandon falsifiability as a useful criterion but rather to attempt to state conditions under which they would be prepared to admit exceptions to the notion of universal multiplicity. For example they recommend the rejection of claims that some discovery, such as microscopy, was already known in the ancient world. On the other hand they also warn against moves by supporters of single discoverers who might impose narrow criteria that exclude potential multiple claims on the basis that they might have been incomplete.

As a further example, it has been claimed that Buffon, Lamarck, Herbert Spencer, and others might have multiplied Darwin's discovery of the theory of evolution. For Lamb and Easton one would be justified in holding that these contributions were indeed incomplete since their accounts lacked any notion of the non-survival of many individuals. The Darwin-Wallace multiple, on the contrary, is far more complete although of course still not identical. Multiples, I suggest, should be viewed in terms of matters of degree rather than as a unitary event which is expected to be identical in every respect. Lamb and Easton end one of their chapters by saying,

> The development of any discovery will be attended by an array of scientists, each having grasped the significance and theoretical appreciation to a greater or lesser degree. It is fortunate, however, that the actual practice of scientists does not strictly conform to abstract and formal criteria and that appeals to the actual history

of science can provide ample evidence in favor of the hypothesis that all scientific discovery and invention is multiple and that singletons are—at least, in principle—multiple.[32]

As I see it, the question of multiple discovery remains seriously underexplored. Rather than trying to explain away this phenomenon, I believe it might well serve to reinforce the proposed view of an essentially organically driven evolution of science by contrast to the prevailing attitude which in many cases amounts to a form of scientism.[33] The more one looks at any development the more one finds intermediate contributions, frequently by names that have completely dropped out of the historical account of all but the most scholarly and detailed histories of science.

"Wrong" ideas can quite frequently lead to progress after they are picked up, modified or transformed by somebody else. This is similar to Pauli's picking up Stoner's work with the third or inner quantum number, or Bohr's seizing on Nicholson's idea to quantize the angular momentum of electrons. Many of Nicholson's ideas seem to have been wrong in retrospect. But why not try to go beyond "right and wrong" and accept a tapestry of gradually evolving ideas, undergoing slight modifications and mutations while collectively producing overall progress for science as a whole.

A NEW COTTAGE INDUSTRY. INCONSISTENCY IN SCIENTIFIC THEORIES

While in the early stages of researching this book I received an invitation to speak at a conference organized by Peter Vickers, a British philosopher of science who works on inconsistency in scientific theories. Vickers recalls how Bohr's calculation of the spectrum of He^+ was remarkably accurate (to 5 significant figures) even though his theory turned out to be false in so many respects. Similarly, Vickers asks how Sommerfeld's formula for the fine structure in the spectrum of atomic hydrogen could be remarkably accurate even though it was operating within the old quantum theory with all the limitations that belonged to that theory.[34] Nevertheless as Vickers writes,

> Introducing the two quantum numbers n and k we find that different orbits which before had the same energy (same n, different k) now have slightly different energies. It is these very slightly different energies which explain the very closely grouped spectral lines which we call the "fine-structure". With these assumptions Sommerfeld is led to the following formula for the allowed energies of the hydrogen atom:

$$E(n,k) = \frac{-RhcZ^2}{n^2}\left(1 + \frac{a^2Z^2}{n^2}\left(\frac{n}{k} - \frac{3}{4}\right)\right) + \ldots$$

> Here c is the speed of light, $\alpha \approx 1/137$, and the dots stand for negligible terms.

Exactly the same formula occurs in the fully quantum mechanical version but now electron spin plays a large role.

Not unreasonably, Vickers asks how Sommerfeld could have arrived at the correct formula without knowing about electron spin and all aspects of quantum mechanics as opposed to the old quantum theory. Could it be because some of his views were correct? Could it be that Sommerfeld's core ideas are correct and that's what brought him temporary success? Similarly, in my case, we might presume that some aspect of Nicholson's theory was latching on to "the truth."

There is much discussion of these kinds of issues in the literature, in connection with the realism and antirealism debate. The clarion call of realists in this context has become the phrase "Divide and Conquer" (*Divide et Impera*). The general idea is that by separating out the correct, from the incorrect, parts of a former theory, one can still claim that at least parts of the refuted theory had latched onto the truth. Some of the defenders of this view include Leplin, Kitcher, and Psillos.[35] Meanwhile philosophers including Laudan, Chang, Cordero, Carrier, Lyons, Chakravartty, and Vickers are skeptical of such a strategy for various reasons.[36] In addition, both diachronic and synchronic versions of this view have been discussed.

These debates are usually waged over very successful theories and scientific entities such as Caloric, Phlogiston, and Ether. I suggest that it would be even more difficult to argue for some form of "preservative realism" in a case like Nicholson's proto-atoms and his atomic theory. This is because it is very difficult to find *anything* that gets preserved in the subsequent theory. In any case I will undertake a review of such work in the philosophy of science in this area in order to see to what extent, if any, it might cohere with my own work.

WHAT TO DO?

Another strategy adopted by some realists has been to withdraw to a weak form of realism such as John Worrall's structural realism.[37] Briefly stated, this is the view that although ontological entities postulated in refuted theories do not survive, mathematical structure does survive in many cases. The motto of the structural realist is that we should be realists about mathematical structure but not about the entities in any particular scientific theory.

However even this move is unlikely to succeed in the case of Nicholson, given the degree to which his theory appears to have been simply "wrong" in so many respects. I do not intend to pursue the defense of realism line, or the realism—antirealism issue in general.

My own preference is to take an altogether different approach that is best explained by reference to the work of Thomas Kuhn. Stated simply, where Kuhn sees revolutions, I see continuity and evolution.

In trying to understand priority disputes, I claim that individuals do not matter, and indeed that priority questions do not matter to the progress of science. I take it that science is more about survival of scientific knowledge as a whole not about the frail egos of individual scientists. But of course as individuals we convince ourselves that our own theories and views are all important and that we must be defended at all costs. As I suggested earlier, only science as a whole benefits from such play-acting.

INTERNAL OR EXTERNAL INCONSISTENCIES IN THEORIES

The question of inconsistent theories can be subdivided into at least two categories. First there are frequently internal inconsistencies within a particular theory. For example, many authors have analyzed Bohr's 1913 trilogy paper and have argued that there are any number of inconsistencies within the four or five central postulates that Bohr made at various stages in the development of his epoch-making theory.[38] Consider how Von Laue reacted soon after Bohr published his theory:[39] "This is nonsense! Maxwell's equations are valid under all circumstances, an electron in an orbit must radiate."

My own belief is that it is impossible for theories to develop without there being some kinds of internal inconsistencies. In fact I would argue that the onus is on authors who wish to argue that consistency is the rule rather than the exception.

Many questions of a general kind can be raised on this subject. For example, just what is the role of logic, which represents an obvious aspect of consistency, in the evolution of a scientific theory? My own inclination is toward the view that logic serves to analyze theories after the facts, but that it cannot hope to capture the process of theories in motion, meaning while they are in the course of development. It is precisely the way that theories develop that interests me.

Nevertheless, a good deal of my emphasis is on examining how successive theories relate to each other and with the known empirical facts. I take this to mean the external consistency of theories rather than their internal consistency. As in the case of internal consistency, I will argue that inconsistency between successive theories is the rule rather than the exception and similarly that some inconsistencies of any particular theory with the facts is a perfectly natural feature of all past and present theories. The fact remains that a number of thoroughly "inconsistent" theories, such as Nicholson's, have been hugely productive and have been the stepping-stones to modern theories that show far greater consistency with the facts.

However, I cannot help wondering whether philosophers like Vickers might be asking the wrong question. The mode in which scientific theories develop does not need to be consistent. The lack of consistency that many authors perceive is not something to be explained away as though it were an anomaly. It is rather an important hint as to the nature of scientific development. I believe that theoretical inconsistency should be taken at face value and perhaps embraced for what it is.

EVOLUTIONARY THEORIES OF SCIENTIFIC DEVELOPMENT

I am advocating an evolutionary theory of the development of science in a fairly literal biological sense that I have already alluded to. Of course evolutionary epistemology comes in many different forms. There is a huge literature on this topic, which means different things to different people, including, among many others, philosophers Donald Campbell and Karl Popper.[40] It will be necessary to review some of this work in order to situate my own claims, which I believe to be rather different in spirit, even if there may be agreement in some respects.

I have claimed that scientific progress does not consist in theories being right or wrong, just as biological evolution is neither right nor wrong. As I see it, biological developments are either suited to their environment or not. Those that are suited are perpetuated in future generations. So, I claim, it is with the development of science. Some ideas are perpetuated, not because they are right or truthful but because they facilitate the progress of scientific knowledge. To put words into Kuhn's mouth, knowledge develops from within the body-scientific rather than being pulled toward an external truth.[41]

Animals are known to evolve in order to adapt to their environments. Furthermore, environments are bound to change with the passage of time. Similarly, I suggest, scientific theories evolve in order to adapt to the particular times that they exist in, rather than in order to conform to some objective or "out there" criteria of eternal truth. To the extent that one can speak of theories describing the "truth" it would have to be that theories provide the best description of the world as it happens to exist at a particular point in time.

As I also suggested earlier, scientific development or growth of knowledge should not be approached via individual discoverers, or through individual theories, but as one essentially undivided entity. But to arrive at such a view one has no option but to examine some individual contributions as I have been doing in this book. Naturally many people have realized the cooperative, trans-individual nature of scientific growth even if it has not been put in quite the terms that I am proposing. It could be argued that such approaches as the Strong Program, Science Studies, Sociology of Science, and other similar programs are all aimed toward elucidating precisely this aspect of science.[42]

But studying the whole of science is not the same as studying the social factors that are supposed to determine scientific discoveries. Here is where I differ from the proponents of these various sociological approaches, as I will attempt to explain. I have not put any attention on the social factors involved in the work of my seven scientists, apart from mentioning a few biographical facts about each one of them. I have rather concentrated on the internal scientific developments with which each one was involved.

REVOLUTIONS OR NOT?

To put the matter simply, I think that Kuhn might have vastly overemphasized the role of scientific revolutions. I also don't believe that an evolutionary view can be compatible with Kuhnian discontinuities. I will return to this question, especially as Kuhn's own work on historical episodes sometimes points to continuity rather than revolution, as in the case of quantum theory.[43]

There is a big literature on the question of whether revolutions truly occur in science. Let me briefly sample what other scholars have written on the subject. First there were a number of critiques, most notably by Stephen Toulmin[44] and D. Shapere,[45] aimed at Kuhn's view that appeared immediately following the publication of his famous book.

Rather than rehashing their arguments I prefer to sample a few contemporary commentaries on Kuhn. Given everything that I have said about the body-scientific, multiple discoveries, and so on, it is essential for me to avoid claiming any originality in saying that there are no real scientific revolutions. Moreover, it is not my primary aim to mount a critique of Kuhn on the question of whether or not scientific revolutions occur, but I still feel somewhat duty bound to entertain such questions as a means of better characterizing my own evolutionary and gradualist position.

For example, the Finnish philosopher Jouni-Matti Kuukkanen has written extensively on Kuhn and has made a special study of his changing views of the notions of revolution. Here are some especially apt passages from an article by Kuukkanen:

> The early Kuhn's wholesale and psychologically drastic revolution becomes a gradual and piecemeal communitarian evolution in the later Kuhn, something that may show simultaneous continuity and discontinuity between prerevolutionary and revolutionary stages.[46]
>
> The older sense of revolution included the idea of widely encompassing transformations, where large bodies of knowledge or entire research programs and orientation are abandoned when scientists jump on the board of a new, more successful paradigm. Second, the old sense of revolution was inherently associated with discontinuity, because the old paradigm is abandoned *in toto* and left in the dust of history, to be possibly returned to only at some unspecified future date.[47]

According to Kuukkanen, the later Kuhn wanted to set the record straight and to admit his earlier mistakes. Kuhn retracted the notion of a gestalt switch and pointed out that it is not a fitting description of scientific revolution, because the entity that undergoes revolution is a community. Meanwhile, a true "gestalt switch" is an individual psychological concept and does not apply to sociological entities. "Unlike the revolutions of the younger Kuhn, there may be continuity between pre- and post-revolutionary periods, and revolutions tend to be localized."[48]

Another philosopher, Vincenzo Politi, writes,[49]

> It remains to be seen, of course, whether Kuhn actually defended the idea that scientific revolutions are "sudden epistemic breaks.

Kuhn is very clear when he says that the Copernican Revolution took a couple of centuries to be completed. In his book *The Black-Body Theory and Quantum Discontinuity, 1894–1912* he also speaks about the conceptual change operated by Planck, but which Planck himself was not able to fully understand. It was only the following generation who appreciated the revolutionary nature of Planck's idea, while Planck himself still did not see any rupture with classical mechanics (Politi, 2015)."

There is now even some doubt over the period in the early modern era commonly known as *the* scientific revolution. According to Dan Garber, writing in one of several recent books intended to celebrate the 50th anniversary of the publication of Kuhn's book.

What we can see in the seventeenth century is the early history of what has come to be called among contemporary philosophers of science the disunity of science, the idea of the scientific enterprise as a bundle of competing programs with different methodological, theoretical, practical bases, in Kuhnian terms, competing paradigms that never fully resolve.

In this way, the period that is generally called the Scientific Revolution looks less like a real revolution, an old regime that enters into crisis before being replaced by a new regime, and more like the Protestant Reformation that happened at roughly the same time. Luther and Calvin challenged the Roman Church, and established churches of their own. But they didn't succeed in replacing the Roman Church with a new and reformed church. Indeed, they didn't form a unified opposition to the Catholic Church either: there isn't anything that you can call *the* Protestant Church.[50]

In her 2010 book *Creating Scientific Concepts*, philosopher Nancy Nercessian has developed what she terms a "cognitive history of science" whereby "paradigm shifts" are gradual and caused by incremental conceptual change, so there are no such things as revolution taking place as "sudden epistemic breaks."[51]

Science historian Dan Garber writes,

> The main example that I have been exploring has been the Scientific Revolution of the sixteenth and seventeenth centuries. It is very interesting that nowhere in *Structure* does Kuhn ever as much as mention this as an example of what he is talking about. In general, the examples he gives in *Structure* are for somewhat narrower revolutions, like the Copernican revolution, or the Newtonian revolution, or the Darwinian revolution, or the Einsteinian revolution. To that extent it might not be that important (or surprising) to Kuhn to discover that the so-called Scientific Revolution wasn't a revolution.[52]

Philosopher Alexander Ehmann has this to say in response to my question of whether there have been any devastating critiques of Kuhn's position on revolutions in science:

> The "scientific revolutionist" can easily refer to a single event, person, group, theory or experiment in the history of science in order to make his point, the "scientific evolutionist" has to oversee much larger time frames, groups of interacting people over time, series of experiments, the development of theories on a large scale, etc. The revolution always happens "now." If it doesn't, it's not a revolution. In contrast, evolutionary processes can't be pinned down to one person or event at a specific time. Their stories are of greater complexity, it takes longer to tell them. They are not punchy, and therefore not devastating.[53]

MORE ON MY EVOLUTIONARY VIEW
OF SCIENTIFIC DEVELOPMENT

Let me return briefly to the minor scientific figures discussed in this book. Nicholson and many minor scientists like him contributed very significantly to the development of early atomic physics. Nicholson, I maintain, was not simply wrong. Rather, he helped Bohr to get started with the program of the quantization of angular momentum. In a sense, Nicholson is as much part of the history as Bohr. But Kuhn's early focus on revolutions inevitably serves to diminish the importance of such marginal figures as Nicholson.

In my view there was no sharp revolution, only an evolution in the development of quantum mechanics. Moreover, this evolution is easier to see from the wider perspective of science as one unified whole than when seen from the contributions of individual scientists or theories. I consider that viewing theory change as *revolutionary* may mask the essentially biological-like growth of science that I am defending in this book.

I submit that evolution drives biological development and ultimately even the way in which we think and develop scientific theories and experimentation. Surely, the unsurprising conclusion from this line of thought must be that all the knowledge of the natural world that we have is ultimately determined by evolutionary biology.[54]

It is important to recall that it is scientific *knowledge* that philosophers of science are trying to elucidate and not the actual way the world is. Scientific knowledge is never right or wrong, because it is not proceeding toward an external truth. It is driven from within, essentially by evolutionary forces, which look back to past science. As is true of many who came before me, including Kuhn, I do not believe that science is directed at some "truth" that exists externally or "out there."

What I wish to add to this view is that science progress is far more organic than usually supposed. By regarding science in this manner we can make better sense of work that—like Nicholson's—succeeded in contributing to the growth of science. And perhaps we can even move beyond such language and ways of thinking as "right" and "wrong" when examining the growth and development of science.

BACK TO KUHN

I now turn to something that I believe Kuhn did recognize correctly. One of my main sources of information on this question has been a recent book by philosopher Brad Wray who claims that in fact Kuhn's later work was primarily aimed toward developing an *evolutionary* epistemology.[55] I want to examine the extent if any to which this aspect of Kuhn's thinking coincides with my own.

> As Kuhn developed his epistemology of science, he saw more and more similarities between biological evolution and scientific change. Consequently as he developed his epistemology of science it became a more thoroughly evolutionary epistemology of science.[56]

As Wray also describes, Kuhn was one of the key philosophers of science who initiated the *historical turn* in the philosophy of science in the early 1960s whereby greater attention was given to the history of science. Although he later changed his attitude, Kuhn came to adopt what he later called a *historical perspective*. This developmental view, as he sometimes called it, is an *evolutionary perspective* on science. More importantly, writes Wray, Kuhn's historical perspective causes us to rethink the role that truth plays in explaining the success of science.

According to Kuhn, we can make better sense of scientific inquiry and the success of science if we see scientific inquiry as "pushed from behind," rather than as aiming toward a fixed goal set by nature. This is not to say that the world does not constrain our theorizing. Kuhn certainly believes it does. But Kuhn wants us to see that the scope of our theories is not determined by nature in advance of our inquiring about them.

Kuhn's first thoughts on epistemology based on evolutionary lines first appear at the end of his book, *The Structure of Scientific Revolutions.*[57] There Kuhn writes that, just as evolution lacks a *telos* and is not driven toward a set goal in advance, so science is not aiming at a goal set by nature in advance. It appears that Kuhn continued to regard this as important to the end of his life. Whatever else he changed, Kuhn did not change this particular aspect of his thinking.

In order to better explain this point Kuhn also claims that, in the history of astronomy, the earth-centered models held the field back for many years. Similarly Kuhn claims, the current truth-centered models of scientific change are holding back philosophy of science, a view with which I concur completely. Somewhat grandiosely, Kuhn also notes a similarity between the reception of Darwin's theory and the reception of his own view on the development of science: Kuhn points out that both views meet the greatest resistance on the claimed elimination of teleology.

But if science is not driven toward the truth how does one explain the success of science? Initially Kuhn did not have a positive answer to this or similar questions. He only argued that it was not a process of marching *toward* the "truth." However, Kuhn later proposed that scientific specialization was the missing positive answer, and that specialization allows scientists to develop more precise conceptual tools for modeling the parts of nature they seek to understand. According to this perspective, just as biological evolution leads to an increasing variety of species so the evolution of science leads to an increasing variety of scientific sub disciplines and specializations. Changes in science are therefore best understood as responses to existing problems, not as attempts to get at the world as it *really is*.

One can give any number of examples to support Kuhn's claim. Out of physics and chemistry there emerges physical chemistry. Biology and chemistry give rise to biochemistry. Biochemistry in turn gives rise to physical-biochemistry and so on. Here is what the philosopher Kuukkanen has to say about Kuhn's evolutionary epistemology (quoted in *The Structure of Scientific Revolutions* [SSR]):[58]

> Although the concept of evolution played a role in Kuhn's thinking from the early stages of his career, the later Kuhn took even greater interest in it. The fact that people hadn't ceased viewing science getting closer to something and begun to see it moving away from something troubled him still in the final stages of his life. (SSR, 307–308, p. 134)
>
> The problem for the early Kuhn was that, although he saw the evolutionary analogy as "nearly perfect" he confesses not being able to "specify in any detail the consequences of this alternative view of scientific advance. (SSR, p. 171),
>
> Further, I will examine to what degree our view of Kuhn's philosophy is altered, if the concept of evolution is taken seriously. It is likely that all who have been accustomed to viewing Kuhn as a philosopher of radical scientific revolutions will be surprised. (SSR, p. 134)
>
> The later Kuhn felt that his evolutionary image of science did not get the amount of attention that it deserved. In his last interview, Kuhn deplored this situation. "I would now argue very strongly that the Darwinian metaphor at the end of the book is right and should have been taken more seriously than it was." (SSR, 307), quote is from p. 134.

REVOLUTION AND EVOLUTION, CAN KUHN HAVE IT BOTH WAYS?

Considering the overall trajectory of Kuhn's work, and the way that scholars such as Brad Wray have characterized it, one might well ask whether Kuhn's revolutionary view is compatible with his later evolutionary view. Personally I do not believe that these two views are compatible and this is where I begin to part company with the excellent writings of Wray and Kuhn himself.

First I want to inquire into why it is that Kuhn's view has been called a social epistemology. Kuhn's own reply would be that science is a complex social activity and that the unit of explanation is the group, not the individual scientist, something that resonates well with my own view, to which I alluded earlier. For Kuhn the growth of science is not successfully tracked by considering individual scientists or individual theories. Consequently Kuhn asks us to judge changes in theory from the perspective of the research community rather than that of the individual scientists involved. Among other things he claims that the early converts to a theory, as well as the holdouts, aid the community in making the rational choice between competing theories.

Presumably Kuhn would have no problem in regarding the work of Nicholson in this way, that is as an early convert or even as a step toward the new theory in the social context of Bohr and others. The more I have worked on the current project the more I have started to see great merit in Kuhn's ideas, or at least some of them. I have also begun to realize that my previous resistance to the Kuhnian view was too stuck on the cartoon Kuhn or "best-seller Kuhn,"[59] as some call it. This is the Kuhn who is supposed to deny progress and who is often taken to be at the root of all evils such as Science Wars and the Sociological turn in general.[60]

But my recent rapprochement with Kuhn has only occurred because I arrived at the idea of an evolutionary epistemology through my own work in asking how a "wrong" theory can be so successful in many cases. Moreover, a number of Kuhn scholars including Wray, Marcum,[61] Gattei,[62] and Kuukkanen have understood Kuhn in a very different way than he is popularly perceived. Kuhn abandoned his early ideas on scientific revolution and replaced them with thoughts on the way that scientists use language and talk of the emergence of new disciplines and sub disciplines to explain how science can make progress.

Even more pertinent to my own view is a passage from Marcum's book on Kuhn,[63]

> Contra logical positivism, for Kuhn the justification of scientific knowledge, especially with respect to its advancement, is not logical but rather organic-one particularly based on competition and selection.

CAN KUHN HAVE HIS CAKE AND EAT IT?

It would appear that at different times in his career Kuhn stressed both revolutions and an evolutionary view of the development of science. But can revolution coexist with evolution in science, as Kuhn seems to believe? First there is the qualification that for the later Kuhn, revolutions are no longer paradigm changes. They are taxonomic or lexical changes. One question that arises is whether this means that the revolutions he initially gave as examples should no longer count as "true revolutions."

According to Brad Wray, revolutions are essential to Kuhn because they are incompatible with the view that scientific knowledge is cumulative and that scientists are constantly marching ever closer to the truth. I disagree with this position. It may well be that scientists are not moving toward a fixed external truth but the development of science may still be gradual rather than revolutionary. After all, biological evolution is not teleological but is nonetheless gradual unless one subscribes to a form of punctuated equilibrium.[64] For me, the main insight from Kuhn is his evolutionary epistemology not his discontinuous view of theory change.

I have many concerns regarding the history of science as consisting of a series of true revolutions. For example, why consider the quantum revolution to have ended in 1912, as Kuhn seems to do? Surely an equally important revolutionary break was the one that Bohr *began* around 1912.[65] Or was the true revolution the coming of quantum mechanics à la Heisenberg and Schrödinger between 1925 and 1926? Or maybe it should be situated with the coming of Quantum Electrodynamics sometime later in late 1940s or even Quantum Chromodynamics in the 1970s? If there are so many revolutions, could it be that the very concept of a revolution in science may cease to be relevant? Evolution, not revolution, is the key feature that drives the growth of science, or so I wish to claim.

TOULMIN

The historian-philosopher Stephen Toulmin is the author of one of the most searing critiques of Kuhn's work that has ever been published. This work is of special interest in the present context, because Toulmin comes to two main conclusions, both of which correspond rather closely to those arrived at independently by me. First, Toulmin is highly critical of Kuhn's claim that scientific change occurs in a revolutionary fashion. Second, Toulmin proposes an evolutionary account of the development of scientific theories.[66]

In his 1972 book, *Human Understanding*, Toulmin opens his section on Kuhn by saying,

> This theory of "intellectual revolutions" accounts for the processes involved in these two kinds of phases in quite different terms: so much so, that the contrast between normal and revolutionary change has acquired something of the same spurious absoluteness as the medieval contrast between rest and motion.[67]

Toulmin adds that the inquiry into how one set of scientific concepts is displaced by another is a project that was undertaken by a number of authors before Kuhn, including Hanson, Merton, Ben-David, and Toulmin himself. This shows that Toulmin has a rather special interest in this topic and perhaps explains the lengths to which he is prepared to take Kuhn to task, and the authority with which he appears to undertake this critique.

As Toulmin writes,

> Kuhn devotes a whole chapter to a discussion of rival paradigms as alternative "world-views." On this fundamental level, a scientific revolution involves a complete change of intellectual clothes. Its effects are so profound that a scientist working under the authority of the new paradigm shares no theoretical concepts with one whose intellectual loyalties are still committed to its predecessor. Lacking a common vocabulary, they can neither communicate with one another about the disagreements, nor formulate common the theoretical topics for discussion and research. Each man will end up by "seeing" the world in ways organized according to his own schema or Gestalt. For what he "sees" when he looks down (say) a microscope will be governed not only by the structure of his eyes and his instruments but also by his particular theoretical paradigm; this will determine what any particular specimen is seen as—whether the scientists concerned will view it as (say) a tissue or globule or vesicular sack or nucleated cell.[68]

Toulmin then turns to his own critique. He asks whether any scientific discipline has ever in fact produced such a radical discontinuity. He wonders whether such a definition as Kuhn's might exaggerate the severity of conceptual changes that actually take place in science. In searching for possible examples of such radical discontinuities Toulmin suggests that we consider the transition from pre-Copernican astronomy to the science of Galileo and Newton, or the topic of Kuhn's first book.[69]

A second candidate for a genuine revolution for Toulmin is the transition between the classical physics of Newton, up to and including Maxwell, to the subsequent relativistic and quantum theories of the 20th century. Again this is a topic that Kuhn has done groundbreaking work on, especially in the case of the quantum theory. Toulmin's response is that neither case represents a fully fledged revolution in the way that Kuhn claims. In the case of Einstein's theory, Toulmin rejects any claim for a revolution by noting that every step of the new theory was considered, discussed, argued over, and scrutinized in great detail. Moreover, Toulmin reminds us that the more profound the theoretical changes that may be proposed, the more prolonged the scientific discussions tend to be.

For Kuhn genuine scientific revolutions lead the two opposing parties to fail to share a common language or any agreed procedure for comparing experimental findings. Toulmin asks us to consider whether the writings of the physicists who lived through these alleged revolutions reveal any such breakdown in communication:

> If there had in fact been any breakdown in communication, of the sort to be expected in in an authentic scientific revolution, we should be able to document it from the testimony of these physicists. What do we find? If it there was such a revolution, the men directly involved were curiously unaware of it. After the event, many of them explained, very articulately the considerations that prompted the decision to switch from a classical to a relativistic position; and they reported these considerations as being the reasons which justified that change, not nearly the motives which caused it. They did not see the switch, in retrospect, merely as an intellectual conversion, to be described by a shoulder-shrug and the disclaimer: "I can no longer see Nature as I did before..." Nor did they treat it as the outcome of nonrational or causal inferences: "Einstein was so very persuasive..." or "I found myself changing without knowing why."[70]

Changes that are proposed by the likes of an Einstein, says Toulmin, are in fact justified by stronger rational reasons than lesser proposals and do not occur in an irrational fashion. Here I would have to side with Kuhn but for rather different reasons. It may seem that the transition occurs in a rational fashion to Toulmin and as an irrational leap to Kuhn. What I propose is a middle position whereby many evolutionary—and perhaps involuntary—changes are thrown up by various members of the scientific community which them interact with each other to finally produce what appears to be Toulmin's reasoned account of a smooth change. In fact, the change consists of a very fine graining of small steps that are performed by numerous players in each field, most of whom are written out of the historical account as I have been arguing.

Toulmin poses the following question to Kuhn and his supporters:

> If there had in fact been any breakdown in communications, of the sort to be expected in a genuine scientific revolution, we should be able to document it from the testimony of these physicists. What do we find? If there was such a revolution, the men directly involved were curiously unaware of it. After the event many of them explained very articulately the considerations that prompted their decisions to switch from a classical to a relativistic position; and they reported these considerations as being the reasons which justified their change, not merely the motives which caused it.[71]

While broadly agreeing with Toulmin I think he may be overemphasizing the role of rationality among these active participants. I suspect that many of them were reading many articles from players of different levels, each of whom was pushing the understanding forward in small incremental steps. Does this represent a fully rational development that Toulmin supports? Does it perhaps speak more to an internal and largely undetected development composed of a multitude of evolutionary experiments conducted by each and every member of the community of scientists who were publishing their views, or perhaps even just expressing them in public lectures? How much of scientific progress can be attributed to sentient and rational beings and how much to the involuntary proliferation of ideas among the members of the scientific community? One is reminded of some of the scientific episodes we have seen in this book involving the likes of Nicholson, van den Broek, Lewis, and the other pioneers of atomic physics and chemistry.

Toulmin then proceeds to broaden what he means by scientific rationality by appealing to the social dimension of science:[72]

> If philosophers of science have been able to, hitherto, to ignore the actual behavior of scientists, in favor of logical questions about their arguments, this is because the intellectual coherence and systematic application of the sciences mark them off so strikingly from the more arbitrary and unmethodical activities of much social life.[73]
>
> Yet a strong case can be made for analyzing the inner structure and empirical relevance of scientific concepts, also, as elements in continuously developing human activities; and for considering their broader significance by seeing how the specific intellectual procedures which are the "micro institutions" of the scientific life are related to the broader professional goals by which the enterprise of science is currently carried forward.[74]

This clearly represents a good beginning but to my mind does not go far enough in the direction of demoting the notion of rationality as a unique and separate all-conquering faculty that lies at the heart of scientific development. Rationality is not separate from everything else that human agents do in the course of their everyday life and which contributes to the growth of science in a tacit fashion.

In the following section I turn to Donald Campbell, an author who is generally regarded as the founder of evolutionary epistemology, although he frequently refused to accept this accolade and named Karl Popper as his true inspiration.

CAMPBELL

Donald T. Campbell was born in Michigan in 1916 into a family of Appalachian Bible Belt church goers. Although Campbell abandoned the church in his high school days he readily admitted to conducting his scholarly work with the zeal of an evangelical preacher. He was initially an undergraduate and graduate student in psychology. He became enamored of evolutionary theory at an early stage of his career and this was to remain as the main ingredient in all of his research. He is often considered the preeminent evolutionary epistemologist and even the consummately argumentative Karl Popper conceded to agreeing with almost everything Campbell had to say on the subject. Cecilia Heyes, the editor of one of many books devoted to Campbell's work, wrote this about Edward Tolman, one of Campbell's early intellectual influences.

> In contrast with other behaviorists of his day, Tolman saw learners—rat and humans—as meandering explorers hoovering up information about their enclosing maze, only a fraction of which would ultimately be useful.[75]

Campbell's other influences included a study of cybernetics, especially the work of W. Ross Ashby on natural selection analogues of learning theory and perception. Campbell's ideas came to the attention of philosophers, including philosophers of science, as a result of his contribution to a compilation of articles on the philosophy of

Karl Popper as edited by Schlipp.[76] During the latter part of his career, Campbell's main focus was on theory of knowledge, or epistemology, to which he gave a thoroughly evolutionary flavor that has had a lasting influence on many fields, not least of which is philosophical naturalism.

However, Campbell was too sophisticated to try to push evolution too far in his view of epistemology. He has been described as having been both a Darwinian and also a social constructivist. The talk of genes as the determining factors in all matters biological was tempered by the belief that genes do not explain such activities as the rise of logic or the development of quantum mechanics in the 20th century. In trying to explain these activities Campbell wore his social constructivist hat, believing that facts, even scientific facts are determined in the context of social organizations. While accepting that individual scientists are objective, Campbell is at pains to point out that it is the collective objectivity of the group that counts the most. He believed in an increased "fit" between systems and their environments by means of "nested hierarchies" of mechanisms which operate via what became a Campbellian slogan of "blind variation and selective retention" or BVSR as it became known in the trade.

According to Heyes,

> Thus the empirical challenge presented by Campbell's selection theory is to identify where in any given system BVSR is occurring, to delineate the entities on which it is based and to model the system-specific mechanisms of proliferation and selective retention.[77]

What I have been doing in this book was not motivated by Campbell's project. What I claim is that my detailed examination of several chemists and physicists of the early part of the 20th century goes some way to illustrating evolutionary epistemology in action. What I am engaged in here is trying to place my own work in the context of what others have done. Given that I have not been part of the evolutionary epistemology cottage industry I have had the immense good fortune of feeling that I have arrived at an original view. It is only in the course of giving public lectures on the subject that various members of the audience have informed me that my views sound a little like X or Y and that I would profit from studying their work, which in most of the cases I will examine I have done for the very first time.

The one exception is the work of Karl Popper, whose influence has had a lasting impact on my own development but not because of what he might have said about evolutionary epistemology. In this respect I have been obliged to return to Popper to see where evolutionary epistemology does or does not fit in with his overall views. Moreover if my own arguments are to make even the slightest contribution, it is rather essential that I explain how my own brand of evolutionary epistemology differs from that of sociologists, psychologists and philosophers who have made it their lifetime's work.

But I am getting sidetracked and should return to Donald Campbell before I digress too far. Campbell like many other twentieth-century thinkers abandoned the classical epistemological stance of basing knowledge on self-evident truths or foundations. Campbell is therefore opposed to the Cartesian program that has been at the heart of so much philosophy, including the logical positivist attempts to secure scientific knowledge in some similarly foundational manner. Campbell like so many evolutionary epistemologists does not regard logic and rationality as the driving force of all knowledge, although he would not want to deny their enormous roles, especially in areas such as science. But for Campbell and other proponents of evolutionary epistemology, the view is one of imperfect knowers. Knowledge, to paraphrase Michael Bradie, is always presumptive, partial, hypothetical, and fallible.[78]

Or as Bradie puts it in the same article, "It is true that the BVSR model claims that, in fact, all expansions of knowledge beyond what is known are groping, blind, stupid, or haphazard...."[79]

Bradie ends this sentence by adding, "but it is not clear why it must be so." Like so many other philosophers who flirt with evolutionary epistemology, Bradie cannot quite climb aboard the Campbellian bandwagon. This talk of groping, blindness, and haphazard stupidity seems to be too much for Bradie to bear. He therefore feels obliged to reject the idea, something that my work has encouraged me to believe might be a small price to pay if the aim is to understand the nature of science. Similarly, Bradie appears to object to Campbell's apparent wish to equate human knowledge with the activities of rats in mazes:

> At the heart of Campbell's view is a metaphysical picture or meta-phor. It is the "insight" that scientists exploring the world are essen-tially no different from rats learning a maze. In this section I want to explore this metaphor and raise some crucial objections to it.[80]

In my own view one should in fact go much further. The so-called foundations of analytical philosophy and the logical positivist school have been based almost entirely in the analysis of logic and language. But these human skills are perhaps too sophisticated, or too recent, to provide a deep enough analysis of how scientific knowledge is acquired and how the scientific community makes progress. For example, I firmly believe that language is a somewhat superficial attri-bute, in a quite literal sense. When we articulate a thought it implies that the thought has developed fully enough to be expressible in lan-guage. Deeper levels of thought, or early impulses which lead us to embark on a particular thought or line of thinking and not couched in language. They are pre-linguistic. These deeper aspects of the human psyche are surely equal contributors to the process of scien-tific exploration, perhaps they are even more significant.

Scientific development for me is far more of a craftlike activity of trying something, perhaps even randomly or haphazardly, and then making small adjustments and seeing where it might lead. I believe that we delude ourselves if we maintain that scientists are thinking rationally all of the time, even if we restrict ourselves to their thinking about strictly scientific issues. This is why I am so taken by the cases that I have examined in this book. None of them knew exactly what they were doing, in a manner of speaking. None of their ideas were thought out in advance. Their crude ideas developed through an evolutionary process which other scientists latched onto and also tweaked in particular ways.

The philosopher of biology Michael Ruse has interesting things to say on the subject of language and its role in the philosophy of science. The following is a quotation taken from an article in the same compilations of papers dedicated to Donald Campbell:

> The traditional philosophical way of attempting to resolve such a dispute as this [the nature of science] is through a priori methods: linguistic analyses for instance. One tries to discover the meaning of words and the ways in which they are used and so forth. One analyzes the concepts involved and sees if they are consistent and the like. What one does not do is turn to the real world in search of answers. Indeed, such a turn is considered not merely unnecessary but slightly vulgar, philosophically speaking. The real task of the philosopher is taken to be that of trying through conceptual analysis to arrive at the truth, and from there of making prescriptions about what the scientist do or should do in a world of perfect rationality.[81]

It is no small wonder therefore that scientists have reacted so negatively to so much that passes for philosophy of science.[82] It has become something of a commonplace to discuss the ways in which the growth of scientific knowledge has caused humans to become more humble. First there was Copernicus who showed us that the earth was not at the center of the universe or even the solar system.

Then came Darwin who discovered that man evolved directly from other animals and thereby banished the view of the superiority of humans and their dominion over the beasts.

What I am proposing, with some trepidation, is that logic and rationality are not the basic governing forces that we usually believe them to be, not even in the pursuit of scientific questions. More important may be urges, hunches, trial and error, serendipity, and just plain trying different things at different times.

Every now and then these kinds of activities produce something that is of use and evolution takes care to nurture these stepping stones to knowledge. But taking on such a view demands a further step in the direction of humility. The modernist dream of the triumph of rationalism is surely dead, and yet philosophers of science still cling to it tenaciously. They are quick to criticize and mock the logical positivists for their foundational pretensions but they still allow logic and language to rule most of their activities. Why, one might ask, has analytical philosophy of science not withered away yet?

Of course some might say that it has withered away and has been replaced by sociological approaches to the study of science. But one has only to think of the recent Science Wars debate to see that the old school insistence on logic and rationality still lives on. It is clear that the analytical school of philosophy is not going to give in without a fight. In addition their opponents, the sociologists and science studies scholars, made it rather easy for them to survive by embracing many forms of relativism which the analytical philosophers can dismiss with the greatest of ease.[83] The fact remains that the sociological school in all its varieties has failed to carry out a close analysis of actual sciences such as chemistry and physics with a few isolated exceptions.

What I have argued for in this book is a form of sociological approach but a radical one. Perhaps I should say a literally social approach, in that I claim that the society of scientists constitute a unified and living organism. But this does not commit me to putting great emphasis on the study of the social conditions that pertained when my subjects were practicing their chemistry and physics. As I see it the lessons to be learned about the growth of science and the way in which the process takes place, are to be found in the science itself. They are to be found among the primary literature of the early twentieth-century atomic scientists who groped around in trying to understand the spectral lines produced by the atoms of the various elements that were subjected to analysis.

Finally, let me return to the question of whether science is unified or not. According to the view I have presented in this book science is fundamentally unified, and more so than is generally believed. I say this with full knowledge of the fact that the failure of reduction has been generally regarded as an indication that science lacks unity. I suggest that any apparent cause for believing in disunity or fragmentation originates with a critique of the logical positivist conception of science and nothing more.[84] I have devoted much time an effort to argue that chemistry does not reduce fully to quantum mechanics but this does not cause me to doubt that there exists an underlying unity or as I have argued, an underlying, organic and tacit unity to the way that science as a whole progresses.[85]

NOTES

1. I discuss a number of such cases in E.R. Scerri, *A Tale of Seven Elements*, Oxford University Press, New York, 2013.

2. E.R. Scerri, "Prediction of the Nature of Hafnium from Chemistry, Bohr's Theory and Quantum Theory," *Annals of Science*, 51, 137–150, 1994; *A Tale of Seven Elements*, Oxford University Press, New York, 2013, chapter 4.

3. E.R. Scerri, *A Tale of Seven Elements*, Oxford University Press, New York, 2013, chapters 4, 5, 6, and 9.

4. E.R. Scerri, "The Discovery of the Periodic Table as a Case of Simultaneous Discovery," *Philosophical Transactions of the Royal Society A*, A 373: 20140172, 2014.

5. J. Levy, *Scientific Feuds*, New Holland Press, London, 2010.

6. R.K. Merton, "Priorities in Scientific Discovery," *American Sociological Review*, 635–659, 1957, quotation is from p. 635.

7. A. Wallace, as quoted in M. Shermer, *In Darwin's Shadow: The Life and Science of Alfred Russel Wallace*, Oxford University Press, New York, 2002, quotation is from p. 292.

8. R.K. Merton, *The Sociology of Science: Theoretical and Empirical Investigations* Chicago University Press, Chicago, 1973, quotation is from p. 293.

9. Things have changed considerably following the advent of the biotechnology industry when academic scientists began to profit enormously from their scientific research.

10. R. Merton, "Priorities in Scientific Discovery," *American Sociological Review*, 635–659, 1957, quotation is from p. 640.

11. E.R. Scerri, *The Periodic Table, Its Story and Its Significance*, Oxford University Press, New York, 2007.

12. A. Gross, "Do Disputes about Priority Tell Us Anything about Science?," *Science in Context*, 11, 161–179, 1998.

13. K. Mendelssohn, *The Quest for Absolute Zero*, World University Press, London, 1966, quotation is from p. 9. This statement follows Mendelssohn's account of how the liquefaction of oxygen was achieved almost simultaneously and independently by Caillet and Pictet.

14. D. Lamb, S.M. Easton, *Multiple Discovery: The Pattern of Scientific Progress*, Avebury, 1984.

15. W. Ogburn, D. Thomas, "Are Inventions Inevitable? A Note on Social Evolution," *Political Science Quarterly*, 37, 1 (March), pp. 83–98, 1922.

16. R. Merton, "Resistance to the Systematic Study of Multiple Discoveries in Science," *Archives of European Sociology*, IV, 237–282, 1963.

17. N. Sarafoglu, M. Kefatos, J.H. Beall, "Simultaneity in the Scientific Enterprise," *Studies in Sociology of Science*, 3, 20–30, 2012.

18. Herbert Simon is another author who, like Lamb and Easton, favors a logic of discovery. Herbert A. Simon, "Does Scientific Discovery Have a Logic?," *Philosophy of Science*, 40, 471–80, 1973.

19. D. Lamb, S.M. Easton, *Multiple Discovery: The Pattern of Scientific Progress*, Avebury, 1984, pp. xiv–xv.

20. Ibid., p. xv.

21. Needless to say, in order to be consistent with my overall position I cannot claim any special originality for the view I am proposing in this book, but only that I may be providing yet another multiple discovery, in this case of the evolutionary and organic nature of scientific knowledge.

22. D. Lamb, S.M. Easton, *Multiple Discovery: The Pattern of Scientific Progress*, Avebury, 1984, p. xvi.

23. R. Merton, "Priorities in Scientific Discovery," *American Sociological Review*, 635–659, 1957.

24. Merton, *The Sociology of Science: Theoretical and Empirical Investigations*, Chicago, University of Chicago Press, 1973, quotation is from p. 356.

25. Ibid., p. 11.

26. As quoted by W. Ogburn, D. Thomas, "Are Inventions Inevitable? A Note on Social Evolution," *Political Science Quarterly*, vol. 37, no. 1 (March), pp. 83–98, 1922, quotation is from p. 88.

27. D. Lamb, S.M. Easton, *Multiple Discovery: The Pattern of Scientific Progress*, Avebury, 1984, quotation is from p. 18.

28. An interesting recent book explores the role of failure in scientific research, S. Firestein, *Failure, Why Science Is So Successful*, Oxford University Press, New York, 2015.

29. Ibid., quotation is from p. 25.

30. Until arriving at this statement I was beginning to think that Lamb and Easton's view was so close to mine as to render my own book redundant.

31. L.W. Alvarez, W. Alvarez, F. Asaro, H.V. Michel, "Extraterrestrial Cause for the Cretaceous–Tertiary Extinction," *Science*, 208 (4448), 1095–1108, 1980.

32. D. Lamb, S.M. Easton, *Multiple Discovery: The Pattern of Scientific Progress*, Avebury, 1984, quotation is from p. 96.

33. A.L. Hughes, "The Folly of Scientism," *The New Atlantis*, 37, 32–50, 2012.

34. Briefly put, the old quantum theory was false in regarding electrons to be moving in deterministic paths; it did not incorporate the wave nature of electrons or any probabilistic notions that were to follow with the advent of quantum mechanics. The old quantum theory also lacked any notion of electron spin, although it was introduced in the days of the transition between this theory and the new quantum mechanics by Pauli. Even at this stage it was not so much the notion of spin, which Pauli initially rejected, that was recognized, but only the need for a fourth degree of freedom to every electron.

35. P. Kitcher, *The Advancement of Science: Science without Legend, Objectivity without Illusions*, Oxford University Press, Oxford, 1993; J. Leplin, *A Novel Defence of Scientific Realism*, Oxford University Press, Oxford, 1997; S. Psillos, *Scientific Realism: How Science Tracks Truth*, Routledge, London, 1999.

36. M. Carrier, "Experimental Success and the Revelation of Reality: The Miracle Argument for Scientific Realism." *In Knowledge and the World: Challenges beyond the Science Wars*, Martin Carrier, Johannes Roggenhofer, Günter Küppers, and Philippe Blanchard, eds., 137–61, Springer, Berlin, 2004; A. Chakravartty, *A Metaphysics for Scientific Realism: Knowing the Unobservable*, Cambridge University Press, Cambridge, 2007; H. Chang, "Preservative Realism and Its Discontents: Revisiting Caloric," *Philosophy of Science* 70, 5, 902–12, 2003; A. Cordero, "Scientific Realism and the Divide et Impera Strategy: The Ether Saga Revisited," *Philosophy of Science* 78, 5, 1120–30, 2011; T.D. Lyons, "Scientific Realism and the Pessimistic Meta-modus Tollens," in *Recent Themes in the Philosophy of Science: Scientific Realism and Commonsense*, Steve Clarke and Timothy D. Lyons, eds., 63–90. Kluwer, Dordrecht, 2002.

37. J. Worrall, "Structural Realism: The Best of Both Worlds?" *Dialectica*, 43: 99–124, 1989.

38. Numerous authors have pointed out the apparent inconsistencies in Bohr's theory of 1913. They include I. Lakatos, "Falsification and the Methodology of Scientific Research Programmes," in I. Lakatos, A. Musgrave, eds., *Criticism and the Growth of Knowledge*, Cambridge University Press, Cambridge, 1970, pp. 91–196; T. Bartelborth, "Is Bohr's Model of the Atom Inconsistent?," in P. Weingartner, G. Schurz, eds., *Philosophy of the Natural Sciences, Proceedings of the 13th International Wittgenstein Symposium (Vienna)*, 1989, 220–223; H. Hettema, "Bohr's Theory of the Atom 1913–1922," *Studies in History and Philosophy of Science*, 26, 307–323, 1995.

39. M. Von Laue, 1914, as cited in M. Jammer, *The Conceptual Development of Quantum Mechanics*, McGraw-Hill, New York, 1966, p. 86.

40. D. Campbell, in P.A. Schlipp, ed., *The Philosophy of Karl Popper*, Open Court, Chicago, 412–463, 1974; K.R. Popper, *Objective Knowledge, An Evolutionary Approach*, Clarendon Press, Oxford, 1972.

41. Kuhn has claimed precisely this, apart from any talk of the body-scientific.

42. D. Bloor, *Knowledge and Social Imagery* (1976); D. Bloor, "The Strengths of the Strong Programme." *Scientific Rationality: The Sociological Turn*, Springer, Netherlands, 1984, pp. 75–94; A. Pickering, ed., *Science as Practice and Culture*, University of Chicago Press, Chicago; H.M. Collins, "Introduction: Stages in the Empirical Programme of Relativism." *Social Studies of Science*, 1981, pp. 3–10.

43. T.S. Kuhn, *The Copernican Revolution*, Harvard University Press, Cambridge, 1957; T.S. Kuhn, *Black-Body Radiation and the Quantum Discontinuity, 1894–1912*, Oxford University Press, Oxford, 1978.

44. S. Toulmin, "Does the Distinction between Normal And Revolutionary Science Hold Water?," in Imre Lakatos, Alan Musgrave, eds., *Criticism and the Growth of Knowledge*, Cambridge University Press, Cambridge, 39–47, 1970.

45. D. Shapere, "The Structure of Scientific Revolutions," *Philosophical Review*, 7, 383–394, 1964.

46. J.-M. Kuukkanen, "Revolution as Evolution," in V. Kindi, T. Abratzis, eds., *Kuhn's The Structure of Scientific Revolutions Revisited*, Routledge, London, 2012.

47. Ibid., p. 138.

48. Ibid.

49. V. Politi, private correspondence. Quoted with permission.

50. D. Garber, "Why the Scientific Revolution Wasn't a Scientific Revolution," in Robert Richards and Lorraine Daston, eds., *Kuhn's Structure of Scientific Revolutions at Fifty*, University of Chicago Press, Chicago, 2016, quotation from p. 142.

51. N. Nercessian, *Creating Scientific Concepts*, MIT Press, Cambridge, MA, 2010.

52. D. Garber, "Why the Scientific Revolution Wasn't a Scientific Revolution," in Robert Richards and Lorraine Daston, eds., *Kuhn's Structure of Scientific Revolutions at Fifty*, University of Chicago Press, Chicago, 2016.

53. A. Ehmann, private correspondence, March 24, 2015. Quoted with permission.

54. The view I present is not without its pitfalls as emphasized by many authors including Natalie Gontier in her article in the *Internet Encyclopedia of Philosophy*, http://www.iep.utm.edu/evo-epis/. Another excellent source on evolutionary epistemology from which I have benefited greatly is W. Callebaut, *Taking the Naturalistic Turn or How Real Philosophy of Science Is Done*, University of Chicago Press, Chicago, 1993.

55. B. Wray, *Kuhn's Evolutionary Social Epistemology*, Cambridge University Press, Cambridge, 2011.

56. Ibid., p. 84.

57. T.S. Kuhn, *The Structure of Scientific Revolutions*, University of Chicago Press, Chicago, 1962, 169–172.

58. J.-M. Kuukkanen, "Revolution as Evolution," in V. Kindi, T. Abratzis, eds., *Kuhn's The Structure of Scientific Revolutions Revisited*, Routledge, London, 2012.

59. I would like to thank Lucia Lewowicz for this very apt phrase.

60. E. von Glasersfeld, *Radical Constructivism: A Way of Knowing and Learning*, Routledge Falmer, London, 1995; A. Kukla, *Social Constructivism and the Philosophy of Science*, Routledge, London, 2000; L. Vygotsky, *Mind in Society*, Harvard University Press, Cambridge, MA, 1978. My own critiques of constructivism in science education appear in, E.R. Scerri, "Constructivism, Relativism and Chemistry," in *Chemical Explanation, Proceedings of New York Academy of Sciences*, vol. 998, J. Earley, ed., New York, 2003, pp. 359–369; E.R. Scerri, "Philosophical Confusion in Chemical Education," *Journal of Chemical Education*, 468–474, 2003.

61. J.A. Marcum, *Thomas Kuhn's Revolutions*, Bloomsbury, London, 2015.

62. S. Gattei, *Thomas Kuhn's "Linguistic Turn" and the Legacy of Logical Empiricism*, Ashgate, Farnham Surrey, UK, 2008.

63. J.A. Marcum, *Thomas Kuhn's Revolutions*, Bloomsbury, London, 2015, quotation is from p. 143.

64. S.J. Gould, *The Structure of Evolutionary Theory*, The Belknap Press of Harvard University Press, Cambridge, MA, 2002; N. Eldredge, *Time Frames: The Evolution of Punctuated Equilibria*, Princeton University Press, Princeton, NJ, 1985.

65. In other places Kuhn considers the advent of the Schrodinger-Heisenberg axiomatic approach to quantum mechanics as the true revolution.

66. I am very grateful to Farsad Mahootian for pointing me in the direction of Toulmin's writings, something he did after I had presented a lecture on John Nicholson at a conference in Indianapolis in December of 2014.

67. S. Toulmin, *Human Understanding: The Collective Use and Evolution of Concepts*, Princeton University Press, Princeton, NJ, 1972, quotation is from p. 98.

68. Ibid., 100–101.

69. T.S. Kuhn, *The Copernican Revolution: Planetary Astronomy in the Development of the Modern World*, Harvard University Press, Cambridge, MA, 1957.

70. S. Toulmin, *Human Understanding: The Collective Use and Evolution of Concepts*, Princeton University Press, Princeton, NJ, 1972, quotation is from p. 104.

71. Ibid., p. 104.

72. Ibid., p. 116.

73. Ibid., p. 142

74. Ibid., pp. 166–167.

75. C. Heyes, in *Selection Theory and Social Construction, The Evolutionary Epistemology of Donald T. Campbell*, C. Heyes, D.L. Hull, eds., State University of New York, Albany, 2001, pp. 2–3.

76. D. Campbell in P.A. Schlipp, ed., *The Philosophy of Karl Popper*, Open Court, Illinois, 412–463, 1974.

77. C. Heyes, "Introduction," in *Selection Theory and Social Construction, The Evolutionary Epistemology of Donald T. Campbell*, C. Heyes, D.L. Hull, eds., State University of New York, Albany, 2001, p. 10.

78. M. Bradie, "The Metaphysical Foundation of Donald Campbell's Selectionist Epistemology." In *Selection Theory and Social Construction, The Evolutionary Epistemology of Donald T. Campbell*, C. Heyes, D.L. Hull, eds., State University of New York, Albany, 2001.

79. Ibid., p. 45.

80. Ibid., p. 46.

81. M. Ruse, "On Being a Philosophical Naturalist, A Tribute to Donald Campbell," in *Selection Theory and Social Construction, the Evolutionary Epistemology of Donald T. Campbell*, C. Heyes, D.L. Hull, eds., State University of New York, Albany, 2001. p. 72–73.

82. I am thinking of the likes of Steven Weinberg, Vic Stenger, and Peter Atkins to name but a few scientists who have stated that philosophy of science contributes nothing to scientific understanding.

83. L.W. Alvarez, W. Alvarez, F. Asaro, H.V. Michel, "Extraterrestrial Cause for the Cretaceous–Tertiary Extinction," *Science*, 208 (4448), 1095–1108, 1980.

84. It has often been said that it is ironic that Kuhn's best-known book should have appeared in a series dedicated to the unity of science given that his own work appears to be critical of any such unity. Perhaps this represents another point of disagreement between myself and Kuhn, although if scientific revolutions do not occur then many of Kuhn's claims such as incommensurability and paradigm shifts also become redundant, along with any threats to the unity of science.

85. For example, a recent book by Bokulich offers some interesting new perspectives on reduction and can help one to maintain a view of unification within science, although I do not have the space to develop such arguments here. A. Bokulich, *Reexamining the Quantum-Classical Relation: Beyond Reductionism and Pluralism*, Cambridge University Press, Cambridge, UK, 2008. Other work with which I have much sympathy include, R. Giere, *Explaining Science: A Cognitive Approch*, University of Chicago Press, Chicago, 1988 and J.J. Seeman, S. Cantrill, *Wrong But Seminal, Nature Chemistry*, 8, 193–200, 2016.

INDEX

Figures and notes are indicated by f and n following the page number.

Abegg, Richard
  biographical information, 63, 64f
  continuation of work by Mendeleev, xxvi
  critiques of, 70, 70f
  disciplinary focus of, xxv
  on electroaffinity, 67f, 68, 70, 77n6
  electrochemistry work of, 64, 67f, 68, 75
  influences on, 66
  as intermediate figure in history of
    science, 76, 77
  on oxidation potentials, 69, 70
  rule of eight formulated by, 75–76
  on valency, 64, 75, 75f
Adiabatic principle, 126–129
Alkali metals, 86, 107
Alpha particles, 43–45, 50–52, 51–52f,
  57, 61n10
Analytical philosophy, xvii–xviii, xix, xxii, 212
Archonium, 38n4
"The Arrangement of Electrons in Atoms
  and Molecules" (Langmuir), 90
Arrhenius, Svante, 63, 66, 75
Ashby, W. Ross, 207
Atkins, Peter, x
"The Atom and the Molecule" (Lewis), 89
Atombau (Sommerfeld), 138

Atomic number
  Bohr on, 43, 56–60
  charge of atoms and, 48, 51, 53–54
  defined, 42
  discovery of, 42, 43, 47
  Moseley on, 15, 43, 56
  Nicholson on, 15
  van den Broek on, 15, 42–43, 49–50,
    54–56, 59f, 60
Atomic weight
  alpha particle scattering and, 51–52, 61n10
  calculation of, 16–23, 18–19f, 39n9
  charge of atoms and, 45, 47, 48, 50
Atomization energy, 69, 70
Atoms. *See also* Atomic number; Atomic weight
  affinity of unlike atoms, 65, 66
  Bohr-Sommerfeld model of, 124, 127, 130
  charge of, 45, 47–48, 50, 51, 53–54
  electronic configuration of. *See* Electronic
    configuration of atoms
  elliptical model of, 123–124, 125
  plum pudding model of, 16, 38n5, 81,
    100n3, 100n5
  proto-atoms, 14–16, 18, 18f
  quantum theory of. *See* Quantum theory
  Aufbauprinzip method, 121, 124, 125, 126

Barkla, Charles, 45, 47, 48
Bent, Henry E., 168n1
Berzelius, Jacob, 65–66
The Black-Body Theory and Quantum Discontinuity (Kuhn), 194
Blind variation and selective retention (BVSR), 208, 209
Bödlander, Guido, 68, 70
Bohr, Niels
 on adiabatic principle, 126, 127–129
 atomic model proposed by, xxi, xxvii, 118, 120, 123–124
 on atomic number, 43, 56–60
 aufbauprinzip method of, 121, 124, 125, 126
 Bury's contributions to work of, xxix
 citation of van den Broek by, 50
 comparison with work of Stoner, 135–138
 continuation of work of, xxv–xxvi
 criticisms of, 104, 105–106, 129–130
 electron configurations by, xxi, 84, 89, 95, 121–123, 122f, 124f
 hafnium, role in discovery of, 98
 internal inconsistencies in scientific theories of, 189
 Nicholson's contributions to work of, xxix, 3, 7–8, 11, 22, 30–35
 noble gas configurations by, 130, 131f
 Pauli on, 125–126
 periodic table as theorized by, xxviii, 119–126, 122f, 124f
 permanence of quantum numbers hypothesis by, 128–129, 141
 precursors to work of, xxi, xxiii
 quantum theory of atom developed by, xxix, 7–8, 50, 84, 120
 spectral phenomena as explained by, 21–22
 on Stoner, 139, 144
 theory of electrons in metals developed by, 39n7
 transition metal configurations by, 135, 135f
Bohrfest lectures, 125
Bohr-Sommerfeld model of atom, 124, 127, 130

Bonds
 chemical, xxvi, 76–77, 79, 101n13
 covalent, 89, 104, 111–112
 ionic, 89, 93, 100n11
 polar, 89
 tetravalent, 137
Bradie, Michael, 209–210
Buckingham, David, xxxiiin1
Buddhism, 182
Burger, Herman C., 133
Bury, Charles
 biographical information, 93–95
 contributions to work of Bohr, xxix
 disciplinary focus of, xxv, 6
 electron configurations by, xxi, 94, 96–97, 97f
 hafnium, role in discovery of, 98–99
 as intermediate figure in history of science, 79
 on Langmuir, 95–96
 on periodic table, 94, 95
 portrait of, 79, 80f
BVSR (blind variation and selective retention), 208, 209

Campbell, Donald T., 191, 207–210
Carrier, M., 188
Cartwright, Nancy, 147n25
Casson, Loic, 151
Chakravartty, A., 188
Chang, H., 188
Chemical bonds, xxvi, 76–77, 79, 101n13
Chemistry
 electron configurations according to, 80, 87–93
 periodic table as viewed in, 42
 relationship with physics, xxviii–xxix
Chemistry & Atomic Structure (Main Smith), 103
Cognitive history of science, 195
Collected Papers of Niels Bohr (Nielsen), 57
Collective unconscious, 182
Copernicus, Nicolaus, 8, 211
Copper, oxidation potential of, 69
Cordero, A., 188

Coronium, 16, 20
Coster, Dirk, 139, 144
Covalent bonds, 89, 104, 111–112
Creating Scientific Concepts (Nercessian), 195
Crookes, William, 14, 165

Dalton, John, 65
Darwin, Charles, xi, xix, xxv, 9, 175, 185, 212
Dauvillier, Alexandre, 105, 148n31
Davies, Mansel, 95, 98
Dawkins, Richard, xxv
De Broglie, Louis, xxxi, 105, 138, 148n31
De Chancourtois, Emile Béguyer, 158
De Hevesy, George, 98–99
De Solla Price, Derek, 182
Dirac, Paul, xxviii, xxxi
Discovery process, 178, 180–182, 184
Drago, Russell S., 113
Dualistic theory of chemical compounds, 66
Dushman, Saul, 97–98

Eastern Philosophy, xxiii–xxiv
Easton, S. M., 179, 180, 181–182, 183–184,
    185–186
Egotism, in priority disputes, 173, 176
Ehmann, Alexander, 195
Ehrenfest, Paul, 126
Einstein, Albert, 204–205
Electroaffinity, 67f, 68, 70, 77n6
Electrochemical series of elements, 66
Electrochemistry, 64–68, 67f, 75
Electrolysis, 65, 66
Electromagnetic theory, 119, 120
Electronegativity, 77–78nn6–7
Electronic configuration of atoms, 81–93
    Bohr on, xxi, 84, 89, 95, 121–123, 122f, 124f
    Bury on, xxi, 94, 96–97, 97f
    in chemistry, 80, 87–93
    historical development of, xx–xxi, xxvii,
        xxix–xxx, 80
    importance of studying, xxvi
    Kossell on, 85–87
    Langmuir on, xxvii, 90–93, 91–93f,
        95–96, 99–100

Lewis on, xxvii, xxix, 87–89, 88–89f
    Madelung rule and, 158–159, 159f, 169n10
    magnetic properties and, 133–135,
        134–135f
    Main Smith on, xxi, 104–105, 107–109, 108f
        for noble gases, 130, 131f
        periodic table and, xxiii, xxx, 123
    in physics, 80, 81–87
    planetary model of, 13–14, 16
    Rutherford on, xxvii, xxix, 89
    Stoner on, xxi, 119, 130–138, 132f,
        134f, 136–137f
    Thomson on, 81–84, 81f
Elementary quantum of action, 120
Elements. *See also* Periodic table;
    specific elements
    atomic weights of, 16–23, 18–19f
    as basic vs. simple substances, 162–164
    criteria for identifying, 51
    dualistic nature of, 163
    electrochemical series of, 66
    half-cell potentials of, 68
    Mendeleev on, 163–164
    in philosophy of science, 162–163, 164
    proto-elements, 14, 15f, 20, 36, 38n4
    rare earth, 55, 98, 110, 146n8
    terrestrial, 14, 18
    transuranic, 99
Elliptical model of atoms, 123–124, 125
Energy, 69, 70, 145n3
Euler, Leonhard, 175–176
Evolutionary philosophy of science
    Campbell on, 207–208, 209–210
    discovery process and, 181, 182
    Kuhn on, 197–199, 200–201
    Popper on, 191, 207, 209
    quantum mechanics and, xxix, xxx, xxxi, 196
    Scerri on, x, xi–xiii, xviii, xxi–xxii,
        4–5, 196–197
    scientific development and, 190–192
Evolutionary realism, 181, 184
Exclusion Principle, xxi, xxix, 133, 141–142,
    144–145, 147–148nn29–30
Exoplanets, xix–xx
External inconsistencies in scientific
    theories, 190

Fermi, Enrico, 99
Fock, Vladimir, xxx
Foundations of Chemistry (journal),
    xvii, xxviii
Fowler, R. H., 119, 133

Gaia hypothesis, xxv, xxxi, 8
Galileo Galilei, 8
Garber, Dan, 195
Gases
    law of combining, 65
    noble, 18, 19f, 85, 87, 130, 131f
Gattei, S., 200
Gay-Lussac, Joseph, 65
Geiger, Hans, 50–53, 59, 60
Geocentric model of solar system, xix
Gestalt switches, 10, 193
Glasstone, Samuel, 99
Goldstücker, Theodore, xxiv
Gradualism, xi
Gross, Alan, 178

Hafnium (element 72), controversy
    regarding discovery of, 98–99, 173, 174
Half-cells, 68–70
Hartree, Douglas, xxx
Heilbron, John, 3, 30–32, 56, 60, 121, 141
Heisenberg, Werner, xxx, 125
Helicoidal systems, 155, 157f
Helium
    alpha particles and, 44
    atomic weight calculation for, 18, 39n9
    controversy surrounding placement in
        periodic table, 164–165
    electronic configurations of, 56, 59, 161
    in left-step periodic table, 153, 155, 156f,
        161–162, 164–165
Heyes, Cecilia, 207, 208
Hicks, William M., 31
Hoyningen-Huene, Paul, xxxii
Human Understanding (Toulmin), 202
Humboldt, Alexander von, 65
Humphry, Davy, 65
Hydration energy, 69

Hydrides, 71, 74, 78n13
Hydrogen
    alpha particles and, 44
    atomic models of, 118, 120, 123–124
    atomic weight calculation for, 16, 18
    Bohr-Sommerfeld model of, 124, 127
    electrochemical nature of, 66
    electronic configuration of, 56,
        59, 71, 73–74
    fine structure formula in spectrum of, 187

Inconsistency in scientific theories,
    186–188, 189–190
Inductive science, xvi, xvii
Inert pair effect, 111–113
Institutional norms, 176–177
Intellectual property rights, 176, 177
Internal inconsistencies in scientific
    theories, 189–190
International Union of Pure and Applied
    Chemistry (IUPAC), 115n13
Ionic bonds, 89, 93, 100n11
Ionic dissociation theory, 66, 68
Ionization energy, 69, 70
Ionization potentials, 113, 113f

Jammer, Max, 32
Janet, Charles
    anticipation of periodic amendments
        by, 110
    on atomic structure and chemistry,
        152–153
    biographical information, 149–151, 150f
    continuation of work by Mendeleev, xxvi
    disciplinary focus of, xxv, 149
    helicoidal systems of, 155, 157f
    historical neglect of, 168
    left-step periodic table by. *See* Left-step
        periodic table
    on Madelung rule, 158–159, 159f
    modern perceptions of work by, 165–166
    modification of periodic table formulated
        by, 166, 167f
Jensen, William B., 75, 77n6

Journal of Chemical Education, 166
Jung, Karl, 182

Kitcher, P., 188
Koestler, Arthur, 182
Kossell, Walther, 85–87, 90, 91, 100*n*11
Kragh, Helge, 34
Kuhn, Thomas
  comparison with work of Scerri,
    12, 200, 201
  critiques of, 202–203
  on discovery process, 178, 182
  evolutionary philosophy of, 197–199,
    200–201
  naturalistic approach of, xvii
  on periodic table theory by Bohr, 121
  on scientific revolutions, xi, xvii–xviii, xxi,
    7, 10, 172, 192–195, 201–205
  on truth and science, 9
Kuukkanen, Jouni-Matti, 193, 199, 200

Lagrange, Joseph-Louis, 176
Lakatos, Imre, 4
Lamb, D., 179, 180, 181–182, 183–184,
  185–186
Landé, Alfred, 130, 141
Langmuir, Irving
  on chemical bonds, 79
  criticisms of, 95–96, 104
  electron configurations by, xxvii, 90–93,
    91–93*f*, 95–96, 99–100
  on periodic table, 92–93, 92*f*
Language, role in philosophy of science,
  xvii–xviii, xix, xxii, 210, 211
Larmor, Joseph, 15, 119, 145
Laudan, Larry, 188
Lavosier, Antoine, 163
Law of combining gases, 65
Left-step periodic table
  in contemporary scholarship,
    159–162, 160*f*
  design of, 150, 169–170*n*11
  helium in, 153, 155, 156*f*, 161–162,
    164–165

proponents of, 168*n*1
quantum mechanics and, 150
variations of, 166, 167*f*
Leplin, J., 188
Leviathan and the Air Pump (Shapin &
  Schaffer), xvii
Levy, Joel, 174
Lewis, Gilbert N.
  on chemical bonds, 76, 77, 79, 101*n*13, 105
  criticisms of, 90, 104
  electron configurations by, xxvii, xxix,
    87–89, 88–89*f*
  octet rule formulated by, 71, 76
Locke, John, 70, 70*f*
Lockyer, Norman, 14
Logic, role in philosophy of science, 4, 212
Logical positivism, 180, 201, 209, 210, 212, 213
The Logic of Scientific Discovery (Popper), 181
Lovelock, James, xxv, xxxi, 8
Lyons, D., 188

Madelung rule, 158–159, 159*f*, 169*n*10
Magnetism
  in electronic configurations, 133–135,
    134–135*f*
  experiments with, 82, 82*f*
  paramagnetism, 134, 134*f*, 135, 138
Main Smith, John D.
  biographical information, 103, 104*f*
  on Bohr, 105–106
  on chemical bonds, 105
  disciplinary focus of, xxv
  electron configurations by, xxi, 104–105,
    107–109, 108*f*
  historical neglect of, 113–114
  inert pair effect and, 111–113
  on Langmuir, 104
  on Lewis, 104
  on periodic table, 106, 109–110, 111
  in priority dispute, 143–144
  on Stoner, 113–114, 115*n*2, 143–144
Marcum, James A., xiii, 200, 201
Marsden, Ernest, 50–53, 59, 60
Mathematical tripos exams, 13, 38*n*1
Mayer, Alfred, 82, 82*f*

Mazurs, E., 165
McCormmach, Russell, 7–8
McMillan, Edwin, 99
Mendeleev, Dimitri
    continuation of work of, xxvi
    on elements as basic substances,
        163–164
    failure to cite work of competitors,
        177–178
    periodic table constructed by, 43, 46, 72f,
        73, 74, 74f
    in priority dispute, 178
    rule of eight formulated by, 64,
        71, 73–75, 76
Mendelssohn, K., 214n13
Merton, Robert, 175–178, 180, 183
Metals
    alkali, 86, 107
    half-cells and, 68–69
    oxidation potentials of, 69
    platinum, 55
    scattering ratios for, 52, 52–53f, 53
    theories of electrons in, 17, 39n7
    transition series of. See Transition metals
Meyer, Lothar, 178
Mingos, Michael, 100–101n13
Moseley, Henry
    on atomic number, 15, 43, 56
    precursors to work of, xxix
    spectroscopic law of, 106
    van den Broek's contributions to work
        of, xxix, 56
Multiple discovery, 178–186
    in history of science, 174, 178–180
    philosophy of science and, 180–182
    priority disputes and, 175, 179
    reasons for, 182–184
    universal multiplicity and, 184–186

Nagaoka, Hantaro, 14, 81, 82
Nationalism in science, 173–174, 177
Naturalism, xvii, 208
Nature (magazine), 19, 20, 47, 50, 53
Nebulium, 16, 20, 25, 30–31
Nercessian, Nancy, 195
Nernst, Walter, 63, 94

Newlands, John, 178
Nicholson, John
    on atomic number, 15
    atomic weights calculated by, 16–23,
        18–19f, 39n9
    biographical information, 13, 14f
    contributions to work of Bohr, xxix, 3,
        7–8, 11, 22, 30–35
    disciplinary focus of, xxv, 6
    as intermediate figure in history of
        science, 5, 37–38
    limitations of theories of, 36–37
    on Orion nebula spectrum, 3, 15, 20,
        23, 23–24f
    Planck's constant and, 27–30
    planetary model of atoms proposed
        by, 13–14, 16
    proto-atoms described by, 14–16, 18, 18f
    proto-elements described by, 14, 15f,
        20, 36, 38n4
    quantization of angular momentum by,
        xxi, xxix, 7, 22, 30, 37–38
    reactions to work of, 30–36, 138
    on solar corona spectrum, 3, 15, 26,
        27, 27–29f
    spectral phenomena as explained by, xii,
        20–22, 21f, 24–26
    theory of electrons in metals developed
        by, 17
Nickel, electronic configuration of,
    92–93, 93f, 100
Nitrogen, electronic configuration of,
    121–122, 122f, 125
Nobel Prize, 79, 99
Noble gases, 18, 19f, 85, 87, 130, 131f
Norms, institutional, 176–177

Octet rule, 71, 76, 79
Ogburn, W., 180
"On the Distribution of Electrons Among
    Atomic Levels" (Stoner), 119
"On the Theory of the Decrease of Velocity of
    Moving Electrified Particles" (Bohr), 57
Orion nebula spectrum, 3, 15, 20, 23, 23–24f
Ornstein, Leonard S., 133
Ostwald, Wilhelm, 63

Oxidation potentials, 69–70
Oxides, 74
Oxygen
electrochemical nature of, 66
electronic configuration of, 71, 73, 74
liquefaction of, 214*n*13

Pais, Abraham, 31, 48
Paramagnetism, 134, 134*f*, 135, 138
Park, Benjamin, 183
Pauli, Wolfgang
on Bohr, 125–126
Exclusion Principle of. *See* Exclusion
Principle
on periodic table, xxviii
precursors to work of, xxi, xxiii
quantum mechanics and, xxxi
Stoner's contributions to work of, xxix,
133, 138, 139, 140–142, 144–145
Pauling, Linus, 77, 77*n*7
Periodic table. *See also* Elements;
specific elements
adiabatic principle and, 126–127
atomic weight calculations for elements in,
18–19, 19*f*
Bohr on, xxviii, 119–126, 122*f*, 124*f*
Bury on, 94, 95
in chemistry, 42
discovery of, 43
electronic configuration of atoms and,
xxiii, xxx, 123
formats for, 100*n*9
Janet's formulation of, 150, 153–155, 154*f*,
156*f*, 159–162, 160*f*
Kossell on, 85–86
Langmuir on, 92–93, 92*f*
left-step model. *See* Left-step periodic table
Main Smith on, 107, 109–110, 111
Mendeleev's construction of, 43, 46, 72*f*,
73, 74, 74*f*
numbering system for groups in, 115*n*13
ordering of, 48
oxidation potentials and trends in, 69–70
Pauli on, xxviii
in physics, 42
theoretical foundations of, ix, x, 150

Thomson on, xxviii, 81, 83
van den Broek's construction of, 42–43,
44–50, 45*f*, 47*f*, 49*f*
Permanence of quantum numbers
hypothesis, 128–129, 141
Perrin, Jean, 13, 81, 82
Philosophical Magazine, 53, 57, 113, 119,
121, 143–144
Philosophy of science
analytical, xvii–xviii, xix, xxii, 212
Eastern Philosophy and, xxiii–xxiv
elements in, 162–163, 164
evolutionary approach to. *See* Evolutionary
philosophy of science
language in, xvii–xviii, xix, xxii, 210, 211
logic and rationality in, 4, 205–206,
210–211, 212
multiple discovery and, 180–182
sociological approach to, xxxi, xxxii, 4,
192, 212–213
Phosphorus, electronic configuration of,
135–136, 136–137*f*
Physics
electron configurations according
to, 80, 81–87
periodic table as viewed in, 42
relationship with chemistry,
xxviii–xxix
Plagiarism, 179, 180
Planck, Max, 119–120, 145*n*3, 194
Planck's constant, 27–30
Planetary model of atoms, 13–14, 16
Platinum metals, 55
Plum pudding atomic model, 16, 38*n*5, 81,
100*n*3, 100*n*5
Polar bonds, 89
Politi, Vincenzo, 193–194
Popper, Karl
on discovery process, 180–181, 182
evolutionary philosophy of, 191,
207, 209
on inductive science, xvi, xvii
Scerri's meeting with, xxxiii*n*3
on scientific revolutions, xi, 10
Post, Heinz, xvi
Potassium, electrochemical nature of, 66
Prigogine, Ilya, 33, 40*n*28

Priority disputes, 98–99, 143–144,
173–178, 179, 189
Proto-atoms, 14–16, 18, 18f
Proto-elements, 14, 15f, 20, 36, 38n4
Proto-fluorine, 16, 26, 27, 29–30
Proto-hydrogen, 38n4, 39n8
Prout, William, 44
Psillos, S., 188

Quantization of angular momentum, xxi,
xxix, 7, 22, 30, 37–38, 146n5
Quantization of energy, 145n3
Quantum mechanics
accuracy of calculations in, xxx
on behavior of matter and radiation, 9
evolutionary development of, xxix,
xxx, xxxi, 196
left-step periodic table and, 150
Quantum theory
adiabatic principle and, 126
Bohr's formulation of, xxix, 7–8,
50, 84, 120
old version of, xxx, 187, 215n34
origins of, 50
Stoner on, 118

Rare earth elements, 55, 98, 110, 146n8
Rationality, role in philosophy of science, 4,
205–206, 210–211, 212
Realism, 181, 184, 187–188
Reduction potentials, 78n8
Rosenfeld, Leon, 32–33, 138
Rule of eight, 64, 71, 73–76
Ruse, Michael, 211
Rutherford, Ernest
on alpha particles, 43–44, 50
atomic model proposed by, 14,
16, 83, 119
on charge of atoms, 45, 47, 48
electron configurations by, xxvii,
xxix, 89
on physics vs. chemistry, xxviii
on plum pudding atomic model, 100n5
Stoner on, 118

Scerri, Eric
blog written by, xx–xxii
comparison with work of Kuhn,
12, 200, 201
on Eastern Philosophy, xxiii–xxiv
educational background, xv–xvii
electron configurations studied by,
xxvi–xxvii
evolutionary philosophy of, x, xi–xiii,
xviii, xxi–xxii, 4–5, 196–197
influences on, xvi, xvii, xviii, xxxiii$n$1
journal founded by, xvii, xxviii
meeting with Popper, xxxiii$n$3
on modification of Janet's periodic
table, 166, 167f
motivations for writing book, xviii–xix,
xxiii, xxiv–xxv, xxxi, 10
objections to work of, 11
periodic table studied by, ix
on priority disputes, 189
as teacher, xvi, xxvi
Schrödinger, Erwin, xxx, xxxi, 146n5
Science. See also Chemistry; Philosophy of
science; Physics
cognitive history of, 195
collective nature of, 5, 8, 37, 181
discovery process in, 178, 180–182, 184
inconsistency in theories of, 186–188,
189–190
inductive, xvi, xvii
intellectual property rights in, 176, 177
marginal and intermediate figures in
history of, 5–6
multiple discovery in. See Multiple
discovery
nationalist influences on, 173–174, 177
priority disputes in, 98–99, 143–144,
173–178, 179, 189
revolutions in, xi, xvii–xviii, xxi, 7, 10, 172,
192–195, 201–205
right vs. wrong ideas in, xviii–xix, 7–11, 22,
109, 186, 197
social dimension of, 206
strengths of, x
Science Wars, xvii, xxii, xxxi, 200, 212
Scientific community, xiii, 8, 9, 22–23, 210

Scientific development
  cultural factors involved in, 184
  evolutionary theories of, 190–192
  impersonal view of, 8
  organic nature of, 9–10, 60, 184, 213
  rationality and, 206, 210–211
SciGaia hypothesis, 8, 9–10
Seaborg, Glen, 110, 153
Server, Daniel, 147n25
Shapere, D., 192
Sidgwick, Neville, 95, 111, 112
Simmons, L. M., 165
Simon, Herbert, 214n18
Simultaneous discovery. *See* Multiple
  discovery
Social constructivism, 208
Sociological philosophy of science, xxxi,
  xxxii, 4, 192, 212–213
Sociology of Science (Merton), 180
Solar corona spectrum, 3, 15, 26, 27,
  27–29f
Solar system, geocentric model of, xix
Sommerfeld, Arnold, 123–124, 125, 134,
  138, 143, 187
Spectral phenomena
  Bohr on, 21–22
  Nicholson on, xii, 20–22, 21f, 24–26
  in Orion nebula, 3, 15, 20, 23, 23–24f
  of solar corona, 3, 15, 26, 27, 27–29f
Spectroscopic law, 106
"The Spectrum of Nebulium" (Nicholson),
  25
Spinoza, Baruch, xxiv
Stewart, Philip, 152, 153, 165, 166
Stoner, Edmund Clifton
  biographical information, 117–118, 118f
  Bohr on, 139, 144
  contributions to work of Pauli, xxix, 133,
    138, 139, 140–142, 144–145
  criticisms of Bohr by, 129–130
  disciplinary focus of, xxv, 6
  electron configurations by, xxi, 119,
    130–138, 132f, 134f, 136–137f
  inert pair effect and, 111–112
  as intermediate figure in history of
    science, 5

  on magnetic properties of transition
    metals, 133–135, 134f
  Main Smith on, 113–114, 115n2, 143–144
  in priority dispute, 143–144
  on quantum theory, 118
  reactions to work of, 138–139
  on Rutherford, 118
Structural realism, 188
The Structure of Scientific Revolutions (Kuhn),
  xviii, 7, 172, 195, 198
Sulfur, electronic configuration of, 136–137

A Tale of Seven Elements (Scerri), 6
Taoism, xxiii, xxiv, 182
The Tao of Chemistry (Scerri), xxiv, xxxiiin14
The Tao of Physics (Capra), xxiii
Terrestrial elements, 14, 18
Tetravalent bonds, 137
Thomas, D., 180
Thomson, J. J.
  on alpha particles, 57
  atomic model proposed by, xxvii, 14,
    16, 38n5, 81
  contributions to work of Nicholson,
    15, 16, 20
  electron configurations by, 81–84, 81f
  on periodic table, xxviii, 81, 83
Tolman, Edward, 207
Toulmin, Stephen, 192, 202–207
Transition metals
  electron configurations for, xxvii, 91,
    96–97, 97f, 130–131
  Madelung rule for, 159
  magnetic properties of, 133–135, 134f
Transuranic elements, 99
Tungsten, electronic configuration of, 97

Universal multiplicity, 184–186
Uranium, 18, 44

Valency, 64, 71–76, 72f, 74–75f
"Valency and the Periodic System"
  (Abegg), 64

van den Broek, Anton
  on alpha particles, 44–45
  on atomic number, 15, 42–43, 49–50,
    54–56, 59*f*, 60
  biographical information, 41–42, 42*f*
  on charge of atoms, 47–48, 50, 51, 53–54
  contributions to work of Moseley, xxix, 56
  disciplinary focus of, xxv, 6
  historical neglect of, 42–43, 60
  as intermediate figure in history of
    science, 5, 60, 61
  periodic table constructed by, 42–43,
    44–50, 45*f*, 47*f*, 49*f*
  scattering ratios calculated by, 53, 53–54*f*
Van Spronsen, Jan, 165
van't Hoff, Jacobus, 63

Vickers, Peter, 186–187, 188, 190
von Laue, M., 189

Wallace, Alfred Russell, 175, 185
What Is This Thing Called Science?
  (Chalmers), xvi
Whiddington, Richard, 57
Whiggism, 11, 22, 97
Wilson, William, 31
Worrall, John, 188
Wray, Brad, 197, 199, 200, 201

Zeeman effect, 140, 144, 145, 147*n*26
Zinc, oxidation potential of, 69–70